IntechOpen

Supporting open minds since 2005

Electron Crystallography
http://dx.doi.org/10.5772/intechopen.83201
Edited by Devinder Singh and Simona Condurache-Bota

Contributors
Devinder Singh, Sven Hovmöller, Fang Fang, Sinziana Paduroiu, Dugan Hammock, Klee Irwin, Ouahida Zeghouan, Mohammad Jafari Eskandari, Reza Gostariani, Mohsen Asadi Asadabad, Tatsuhiko Aizawa, Tomomi Shiratori, Takafumi Komatsu, Yohichi Kohzuki, Seifeddine Sellami, Mohamed AbdEsselem Dems

First published in London, United Kingdom, 2020 by IntechOpen
IntechOpen is the global imprint of INTECHOPEN LIMITED, registered in England and Wales, registration number: 11086078, 7th floor, 10 Lower Thames Street, London, EC3R 6AF, United Kingdom
Printed in Croatia

British Library Cataloguing-in-Publication Data
A catalogue record for this book is available from the British Library

Additional hard and PDF copies can be obtained from orders@intechopen.com

Electron Crystallography
Edited by Devinder Singh and Simona Condurache-Bota
p. cm.
Print ISBN 978-1-83880-189-2
Online ISBN 978-1-83880-190-8
eBook (PDF) ISBN 978-1-83968-284-1

For EU product safety concerns:
IN TECH d.o.o., Prolaz Marije Krucifikse Kozulić 3, 51000 Rijeka, Croatia, info@intechopen.com or visit our website at intechopen.com.

Meet the editors

Dr. Devinder Singh is currently working as an assistant professor in the Amity School of Applied Sciences, Amity University, Lucknow, India. He is a specialist in Expt. Condensed Matter Physics/Materials Science with a Ph.D. from the Institute of Science, Banaras Hindu University (B.H.U), Varanasi, India; a post-doctorate degree from Stockholm University, Sweden and then work experience as a DST Inspire Faculty at Panjab University, Chandigarh. He has contributed over 37 peer-reviewed international research publications and authored four book chapters, two book articles and one book. His research work has been selected for talks and discussions in several premier international and national conferences. He has mentored graduate, postgraduate and PhD students. He has been the recipient of several prestigious international/national fellowships and awards such as the Seal of Excellence from European Commission, DST Inspire Faculty Award, Postdoctoral Fellowship etc. He has strong research professional skills in nanomaterials, alloys, structure-property correlations, electron microscopy and metal physics.

Mrs. Simona Condurache-Bota is currently working as an associate professor at The Faculty of Sciences and Environment, Dunarea de Jos University of Galati, Romania. She has post-doctorate degrees in Industrial Engineering and Atmospheric Physics, a Ph.D. in Solid State Physics, and M.Sc. and B.Sc. degrees in Physics, with a M.Sc. scholarship in the Netherlands given by the European Physical Society. She has participated in 17 research and/or educational projects, some of them with national and/or international grants. She has published 28 ISI (Thomson Reuters/Clarivate Analytics) papers, 7 ISI proceedings papers, and 10 books and book chapters. She has over 100 quotations and a Hirsch index of 7. She has participated in over 40 conferences. Her scientific interests are: optics, spectroscopy, semiconducting films, environmental science.

Contents

Preface

In the quantitative determination of new structures, micro-/nano-crystalline materials pose significant challenges in the field of materials, chemical and bio-chemical crystallography. The different properties of materials are structure-dependent. Traditionally, X-ray crystallography has been used for the analysis of these materials. Electron crystallography offers an alternative approach in the study of very small crystals that are a million times smaller than those needed for X-ray diffraction. Under certain conditions, electron crystallography has some advantages over X-ray crystallography: it is possible to record single crystal information from individual micro-/nano crystals and sometimes it helps in the resolution of overlapped reflections in X-ray powder data. This is possible due to the greater scattering cross section of matter for electrons than X-rays. The defects in crystals can be studied using high resolution electron microscopy images. In recent years, electron diffraction has been widely applied for determining the structure of unknown crystals. With the introduction of advanced methodologies, important methods for crystal structural analysis in the field of electron crystallography have been discovered. There are two methods that are developed for collection of complete three-dimensional electron diffraction data: the rotation electron diffraction (RED) and automated electron diffraction tomography (ADT). Large numbers of crystal structures have been solved using the RED method. These include the most complex zeolites ever solved, open-framework compounds and quasicrystal approximants, such as the pseudo-decagonal approximants.

Chapter 1 is related to the introduction about the electron crystallography. In Chapter 2, electron diffraction, high resolution transmission and scanning transmission electron microscope imaging in materials research, especially in the study of nano-science, are presented. In Chapter 3, a comprehensive review and comparison of different methods used for generating the empires are presented. The focus is given to the 'cut and project method', which can be generalized to calculate empires for any quasicrystals. In Chapter 4, the structure analysis of complex pseudo-decagonal (PD) quasicrystal approximants PD2 and PD1 using the RED method have been discussed. PD2 and PD1 are built of characteristic 2 nm wheel clusters with 5-fold rotational symmetry. In Chapter 5, the crystallographic structure evolution of stainless steels during rolling, piercing and plasma nitriding is described and the mechanism of microstructure evolution during metal forming and materials processing is presented. The EBSD (Electron Back Scattering Diffraction) technique is employed for the crystallographic analysis. In Chapter 6, the experimental and theoretical charge-density analysis has been performed to understand the topological and electrostatic properties of the 1-10 phenanthroline hydrate molecule. In Chapter 7, a study on the interaction of dislocation-dopant ions during plastic deformation by strain-rate cycling tests has been presented and discussed.

We wish to express our sincere appreciation to IntechOpen for giving us the opportunity to publish a book on the topic of electron crystallography. We would like to thank all the authors for their significant contributions and for providing high-quality research to share worldwide. We hope that this book will support current researchers and prove to be very useful to the scientific community. We also want to express our sincere gratitude to Ms. Dolores Kuzelj for her help during the entire publication process.

Dr. Devinder Singh
Amity School of Applied Sciences,
Amity University,
Lucknow, India

Dr. Simona Condurache-Bota
Chemistry, Physics and Environment Department,
Dunarea de Jos University,
Galati, Romania

Chapter 1

Introductory Chapter: Electron Crystallography

Devinder Singh

1. Introduction

The different properties of materials are structure dependent. There are many techniques that are developed for the structure analysis. The most common of them is X-ray crystallography for the structural study of periodic ordered structures at atomic scale level. Few years later, after the discovery of X-ray diffraction, electron diffraction of single crystals was invented. The wave nature of electrons was utilized to discover the state of the art instrument, electron microscope. Since then, electron microscope has been extensively used in many fields for the study of micro-/nanomaterials. Electron crystallography is used to collect different information by electron scattering. This has been used to study crystal structures and defects. After first electron microscopy image taken in 1933, the constant engineering developments from the last 80 years or so made it possible today to record high resolution transmission electron microscopy (HRTEM) images. Moreover, the powerful computers play a very important role in the further improvement of HRTEM images as well as to analyze them quantitatively by using different image processing programs. Modern transmission electron microscope (TEM) can be used for both structure and chemical analysis. The structure analysis is performed by electron diffraction and HRTEM, while the chemical analysis is performed by energy dispersive spectroscopy (EDS) and electron energy loss spectroscopy (EELS).

Although X-ray crystallography is known to be the best technique for the structure determination of unknown crystals, but under certain conditions, electron crystallography has some advantages over X-ray crystallography:

- An electron scatters much more strongly than X-rays. Thus, much smaller crystals, million times smaller than those needed for single crystal X-ray diffraction can be studied by electron crystallography. The structure analysis (such as grain boundaries, phases, etc.) of crystals, too small for X-ray diffraction, can be done by electron crystallography.

- HRTEM images of crystals can be recorded in electron crystallography while in X-ray crystallography, imaging is not possible. The phase information remains preserved in the case of HRTEM images.

- The interaction mechanism with the crystal is also different for electron diffraction and X-ray diffraction. Electrons study electrostatic potential distribution in crystals while X-rays study electron density distribution in crystals.

- The X-ray powder diffraction gives results from a sample which may contain millions of small crystals while electron diffraction gives results from a single or just few crystals. In some materials, where the phases are found to intergrow with each other cannot be solved by X-ray crystallography, whereas in electron crystallography, this problem is eliminated as much smaller crystals are needed for electron diffraction.

- The defects in crystals can be studied using HRTEM images.

Radiation damage is the common problem in X-ray crystallography and electron crystallography. This is especially troublesome in the case of electron crystallography. The proteins and organic molecules are the main sufferers when they are being imaged. Radiation damage can be limited by using electron cryo-microscopy where samples go through cryo-fixation and imaging done at liquid nitrogen or liquid helium temperatures. Due to this, X-ray crystallography is more successful for the structural study of proteins that are more prone to radiation damage. Recently, radiation damage was investigated by MicroED for three-dimensional (3*D*) thin crystals [1, 2]. In the past few years, several protein structures are studied by electron crystallography. The studies on inorganic crystals by electron crystallography were first done by Klug [3] and Hovmöller et al [4]. They used HRTEM images as it is possible to choose thin regions along the edge of the crystal for structure analysis using crystallographic image processing program (CRISP).

In recent years, electron diffraction is widely applied for determining the structure of unknown crystals [5–8]. Electron diffraction is a technique which is well-complementing other techniques, single crystal X-ray diffraction and powder X-ray diffraction for determination of structure. Electron diffraction plays a very important role when crystals are very small for study using single crystal X-ray diffraction or very complex for structure solution by powder X-ray diffraction. The main drawbacks of electron diffraction are the problems in the complete collection of 3*D*-electron diffraction data using standard diffraction techniques and the collection of data is very time consuming. Also, the electron diffraction intensities suffered from dynamical scattering. There are two methods which are developed for the collection of complete 3*D* electron diffraction data: the rotation electron diffraction (RED) and automated electron diffraction tomography (ADT) [8, 9].

With the introduction of advanced methodologies, recently, one of the most important methods for crystal structural analysis in the field of electron crystallography has been discovered. This software-based method is named as rotation electron diffraction (RED), which is capable of overcoming the drawbacks and reducing the dynamical effects [8]. There are two computer programs in a software package, that is, RED data collection and RED data processing. The program for collection of 3*D*-electron diffraction data used the combination of goniometer tilt and electron beam tilt in a transmission electron microscope (TEM). A fine step in the range 0.05–0.20° of electron beam tilt combined with a coarse step in the range 2.0–3.0° of goniometer tilts at common tilt-axis allowed to cover a crystal in a large range of tilt. At every combination, electron diffraction frames are collected. With the collection of about 1000–2000 electron diffraction frames, a complete 3*D* data set is obtained. Thus, a complete 3*D*-electron diffraction data set from a submicrometer sized single crystal can be collected in about 2 hours.

The program processes 3*D*-RED electron diffraction data created by the program RED data collection. It consists of correction of shift in electron diffraction frames, peak search in individual electron diffraction frames for diffraction spots, and identifying diffraction spots as reflections in 3*D*. The program containing RED data processing used to view and analysis of 3*D* reciprocal lattices which are reconstructed from electron diffraction frames. RED method is more capable for determination of structure and identification of phase of crystals which are not known. It is faster, easier, and more straight-forward than powder XRD and other techniques based on electron microscopy. The configuration used in RED is similar to the technique involving rotation as used in X-ray diffraction. Along one rotation axis, the sample is rotated continuously. Instead of doing continuous crystal rotation, coarse crystal rotation and fine electron-beam tilt on the same axis of rotation are combined in TEM. At every combination of crystal rotation and beam tilt, electron diffraction frame is collected. The $3D$-reciprocal lattice is reconstructed from the collected electron diffraction frames with the help of RED data processing program.

RED has been used in the recent years to solve large number of crystal structures. These include the most complex zeolites ever solved, open-framework compounds, and quasicrystal approximants, such as the pseudo-decagonal approximants [10–13]. The quasicrystal approximants are dense intermetallic compounds and consist of heavy elements. Thus, they are usual to have higher dynamical scattering. It is more challenging to solve the structure of quasicrystal approximants using RED. Quasicrystalline phases are having forbidden rotation symmetry in their electron diffraction pattern, which are not compatible with periodic translation. Quasicrystals are discovered in 1982 by D. J. Shechtman. A new idea of ordered but non-periodic arrangement of atoms exhibiting sharp diffraction peaks (with icosahedral symmetry) is created due to the breakthrough experiments by Shechtman et al. [14] on rapidly quenched alloys (AlMn) using TEM. Due to the aperiodicity, these materials can be used for applications in industries as they are having properties different from that of conventional metallic materials [15–24]. Quasicrystals are known by a group of different properties, for example, higher hardness, low friction, low-surface energy, and thermal expansion which is as good as metals [25–29].

After the quasicrystals discovered in rapidly quenched AlMn alloys [14], the most important part of their studies has been related to their structure solution. The structure of quasicrystals has not been yet solved. A number of crystalline phases have been observed which are having crystals made up by the same clusters as found in quasicrystals, and they are named as quasicrystal approximants. These approximants are often found to co-exist with quasicrystals. They have similar electron diffraction patterns and similar chemical compositions as that of quasicrystals [30–35]. The local atomic structures are similar in both quasicrystals and quasicrystal approximants. Thus, the structural study of quasicrystal approximants is an efficient way to understand the quasicrystalline structure. Many quasicrystal approximant series have been observed which are closely related on the basis of their structures. Out of the many quasicrystal approximants that are found till yet, only small number of them has been solved to atomic resolution. In this context, RED method is used to solve two very complicated alloy structures of pseudo-decagonal quasicrystal approximants PD2 and PD1 in Al-Co-Ni alloy system [12, 13]. These are built of characteristic 2 nm wheel clusters with five-fold rotational symmetry [36].

Author details

Devinder Singh
Amity School of Applied Sciences, Amity University, Lucknow Campus, Lucknow, U.P., India

*Address all correspondence to: dsingh2@lko.amity.edu

References

[1] Nannenga BL, Shi D, Leslie Andrew GW, Gonen T. High-resolution structure determination by continuous-rotation data collection in MicroED. Nature Methods. 2014;**11**:927-930

[2] Hattne J, Shi D, Glynn C, Zee CT, Gallagher-Jones M, Martynowycz MW, et al. Analysis of global and site-specific radiation damage in cryo-EM. Structure. 2018;**26**:759-766

[3] Klug A. Image analysis and reconstruction in the electron microscopy of biological macromolecules. Chemica Scripta. 1978/79;**14**:245-256

[4] Hovmöller S, Sjögren A, Farrants G, Sundberg M, Marinder BO. Accurate atomic positions from electron microscopy. Nature. 1984;**311**:238

[5] Zhang D, Oleynikov P, Hovmöller S, Zou XD. Collecting 3D electron diffraction data by the rotation method. Zeitschrift fuer Kristallographie. 2010;**225**:94-102

[6] Zou XD, Hovmöller S, Oleynikov P. Electron crystallography: Electron microscopy and electron diffraction, chapter 11. In: International Union of Crystallography. Oxford, England, UK: Oxford Science Publications; 2011. ISBN: 978-0-19-958020-0

[7] Willhammar T, Sun J, Wan W, Oleynikov P, Zhang D, Zou XD, et al. Structure and catalytic properties of the most complex intergrown zeolite ITQ-39 determined by electron crystallography. Nature Chemistry. 2012;**4**:188-194

[8] Wan W, Sun J, Su J, Hovmöller S, Zou XD. Three-dimensional rotation electron diffraction: Software RED for automated data collection and data processing. Journal of Applied Crystallography. 2013;**46**:1863-1873

[9] Kolb U, Mugnaioli E, Gorelik TE. Automated electron diffraction tomography—A new tool for nano crystal structure analysis. Crystal Research and Technology. 2011;**46**:542-554

[10] Martinez-Franco R, Moliner M, Yun Y, Sun J, Wan W, Zou XD, et al. Synthesis of an extra-large molecular sieve using proton sponges as organic structure-directing agents. Proceedings of the National Academy of Sciences of the United States of America. 2013;**110**:3749-3754

[11] Su J, Kapaca E, Liu L, Georgieva V, Wan W, Sun J, et al. Structure analysis of zeolites by rotation electron diffraction (RED). Microporous and Mesoporous Materials. 2013;**189**:115-125

[12] Singh D, Yun Y, Wan W, Grushko B, Hovmöller S, Zou XD. A complex pseudo-decagonal quasicrystal approximant $Al_{37}(Co,Ni)_{15.5}$ solved by the rotation electron diffraction (RED) method. Journal of Applied Crystallography. 2014;**47**:215

[13] Singh D, Yun Y, Wan W, Grushko B, Hovmöller S, Zou XD. Structure determination of a pseudo-decagonal quasicrystal approximant by the strong-reflections approach and rotation electron diffraction. Journal of Applied Crystallography. 2016;**49**:433-441

[14] Shechtman D, Blech I, Gratias D, Cahn J. Metallic phase with long-range orientational order and no translational symmetry. Physical Review Letters. 1984;**53**:1951-1953

[15] Yadav TP, Singh D, Tiwari RS, Srivastava ON. Enhanced microhardness of mechanically activated carbon–quasicrystal composite. Materials Letters. 2012;**80**:5-8

[16] Singh D, Mandal RK, Tiwari RS, Srivastava ON. Nanoindentation characteristics of $Zr_{69.5}Al_{7.5-x}Ga_xCu_{12}Ni_{11}$ glasses and their nanocomposites. Journal of Alloys and Compounds. 2011;**509**:8657-8663

[17] Singh D, Yadav TP, Mandal RK, Tiwari RS, Srivastava ON. Effect of Ga substitution on the crystallization behaviour and glass forming ability of Zr–Al–Cu–Ni alloys. Materials Science and Engineering A. 2010;**527**:469-473

[18] Singh D, Yadav TP, Mandal RK, Tiwari RS, Srivastava ON. Indentation characteristics of metallic glass and nanoquasicrystal-glass composite in Zr-Al (Ga)-Cu-Ni alloys. Intermetallics. 2010;**18**:2445-2452

[19] Singh D, Yadav TP, Mandal RK, Tiwari RS, Srivastava ON. Effect of Ti addition on the quasicrystalline phase formation and indentation characteristics of $Zr_{69.5}Al_{7.5}Cu_{12}Ni_{11}$ alloy. Philosophical Magazine. 2011;**91**:2837

[20] Singh D, Mandal RK, Tiwari RS, Srivastava ON. Effect of cooling rate on the crystallization and mechanical behaviour of Zr–Ga–Cu–Ni metallic glass composition. Journal of Alloys and Compounds. 2015;**648**:456-462

[21] Singh D, Singh D, Mandal RK, Srivastava ON, Tiwari RS. Glass forming ability, thermal stability and indentation characteristics of Ce75Al25—xGax metallic glasses. Journal of Alloys and Compounds. 2014;**590**:15-20

[22] Singh D, Singh D, Mandal RK, Srivastava ON, Tiwari RS. Effect of quenching rate on the microstructure and mechanical behavior of Ce75Al21Ga4 glassy alloy. Materials Characterization. 2017;**134**:18-24

[23] Singh D, Singh D, Mandal RK, Srivastava ON, Tiwari RS. Effect of annealing on the devitrification behavior and mechanical properties of rapidly quenched Ce-based glassy alloys. Journal of Non-Crystalline Solids. 2016;**445**:53-60

[24] Singh D, Singh D, Tiwari RS. Effect of Ga substitution on low temperature transport and magnetic response of Ce75Al25 metallic glass. AIP Advances. 2018;**8**:095222

[25] Singh D, Singh D, Yadav TP, Mandal RK, Tiwari RS, Srivastava ON. Synthesis and indentation behavior of amorphous and Nanocrystalline phases in rapidly quenched Cu–Ga–Mg–Ti and Cu–Al–Mg–Ti alloys. Metallography Microstructure and Analysis. 2013;**2**:321-327

[26] Singh D, Yadav TP, Mandal RK, TiwariRS,SrivastavaON.Nanoindentation studies of metallic glasses and nanoquasicrystal glass composites in ZrAl(Ga)CuNi alloys. International Journal of Nanoscience. 2011;**10**:929-933

[27] YadavTP,SinghD,ShahiRR,ShazMA, Tiwari RS, Srivastava ON. Formation of quasicrystalline phase in $Al_{70-x}Ga_xPd_{17}Mn_{13}$ alloys. Philosophical Magazine. 2011;**91**:2474

[28] Singh D, Singh D, Mandal RK, Srivastava ON, Tiwari RS. Crystallization behavior and mechanical properties of $(Al_{90}Fe_5Ce_5)_{100-x}Ti_x$ amorphous alloys. Journal of Alloys and Compounds. 2016;**687**:990-998

[29] Singh D, Singh D, Srivastava ON, Tiwari RS. Microstructural effect on the low temperature transport properties of Ce-Al (Ga) metallic glasses. Scripta Materialia. 2016;**118**:24-28

[30] Bancel PA. Dynamical phasons in a perfect quasicrystal. Physical Review Letters. 1989;**63**:2741-2744

[31] Abe E, Tsai A. Quasicrystal-crystal transformation in Zn-Mg–rare-earth alloys. Physical Review Letters. 1999;**83**:753-756

[32] Singh D, Tiwari RS, Srivastava ON. Phase formation in rapidly quenched Cu-based alloys. Journal of Materials Science. 2009;**44**:3883-3888

[33] Zou X, Fung K, Kuo K. Orientation relationship of decagonal quasicrystal and tenfold twins in rapidly cooled Al-Fe alloy. Physical Review B. 1987;**35**:4526-4528

[34] Singh D, Hovmöller S, Grushko B, Wan W, Yun Y, Zou XD. A complex pseudo-decagonal quasicrystal approximant solved by strong reflections approach. Acta Crystallographica. Section A. 2017;**A73**:c1192

[35] Hovmöller S, Singh D, Wan W, Yun Y, Wan W, Grushko B, et al. Quasicrystal approximants solved by rotation electron diffraction (RED). Acta Crystallographica. Section A. 2014;**A70**:c1195

[36] Hovmöller S, Zou L, Zou XD, Grushko B. Structures of pseudo-decagonal approximants in Al–Co–Ni. Philosophical Transactions of the Royal Society London: Series A. 2012;**370**:2949-2959

Chapter 2

Transmission Electron Microscopy of Nanomaterials

Mohammad Jafari Eskandari, Reza Gostariani and Mohsen Asadi Asadabad

Abstract

Structural and analytical characterization, in the nanometer scale, has become very important for all types of materials in recent years. Transmission electron microscope (TEM) is a perfect instrument for this purpose, which is summarized in this chapter. Parameters such as particle size, grain size, lattice type, morphological information, crystallographic details, chemical composition, phase-type, and distribution can be obtained by transmission electron micrographs. Electron diffraction patterns of nanomaterials are also used to acquire quantitative information containing size, phase identification, orientation relationship and crystal defects in the lattice structure, etc. In this chapter, typical electron diffraction, high-resolution transmission and scanning transmission electron microscope imaging in materials research, especially in the study of nanoscience are presented.

Keywords: nanomaterials, characterization, transmission electron microscopy, electron diffraction

1. Introduction

Nanotechnology is considered to be the main technology for all types of materials in the current century. Nowadays, the development and production of nanostructure materials aimed at increasing the strength to weight of structures that led to cost and energy saving were considered by researchers [1, 2]. For studying materials in the nanometer scale, the investigation of nanostructures is needed to discover the properties of nanostructured materials. This purpose will not be achieved except by the use of efficient characterization instruments. Transmission electron microscope (TEM) has evolved over many years into a highly sophisticated instrument that has found widespread application across the scientific disciplines. Due to unparalleled ability to provide structural and chemical information over a range of length scales down to the level of atomic dimensions, TEM has developed into an indispensable tool for understanding the properties of nanostructured materials and in manipulating their behavior. The precise control of nanoparticles size, grain size, size distribution and homogeneity, lattice type, crystal structure, dispersion, chemical and physical property of phases such as number, morphology, and structure of the phases at the nano-level are characterized by TEM. Besides, this investigation attempts to demonstrate the effectiveness of EDP technique to the analysis of nanomaterial properties. Types of diffraction patterns such as ring, spot and Kikuchi patterns, and general and unique indexing diffraction patterns are described.

The methods for indexing simple, complicated and imperfect patterns as well as Kikuchi lines and a combination of Kikuchi lines and spots are determined.

In this research, samples of materials such as nanoparticles, nanotubes, bulk metallic, graphene, graphene oxide, and polymer nanocomposites are investigated using an EM208S (PHILIPS) transmission electron microscopy operating at an accelerating voltage of 100 kV and a digital camera. In addition, electron diffraction pattern of several materials are expounded and structure of materials is predicted with electron diffraction pattern results interpretation. On the other hand, TEM/STEM (STEM, JEOL JEM-2100F) equipped with an energy-dispersive X-ray spectrometry (EDS) and the operating voltage of 200 kV are used to evaluate the in-situ phase characterization, microstructural observation, dislocation pile-up and chemical composition of in-situ nanocomposite fabricated by planetary ball milling of Al/BN composite powders and hot extrusion. This paper discusses the basic principle and applications of the transmission electron microscope (TEM) in the field of nanotechnology research.

2. Transmission electron microscope

2.1 Preferred orientation and texture

In some samples, the preferred orientation of planes creates by several mechanical processes such as accumulative roll bonding (ARB), cyclic extrusion compression (CEC) and equal channel angular pressing (ECAP) processes, etc. TEM image and SAED pattern of nanostructured Al 2024 alloy under ECAP process were demonstrated in **Figure 1(a)** and **(b)**, respectively. The SAED pattern of preferred orientation and textured nanostructured materials was formed from many partial rings or incomplete rings. Diffraction pattern of such materials can be described as an intermediate sample between a single and polycrystalline material. The texture produced in materials can be studied by the interpretation of their SAED. The preferred orientation {110} [001] was formed in Al 2024 alloy by this process [3].

2.2 Dislocation

ECAP [3] and cryo-cross-rolling (CCR) process [4, 5] were carried out on Al 5083 and Al 1050 alloys, respectively. These two processes form a matrix with highly dislocation density, which changes the diffraction pattern. Besides, SAED patterns

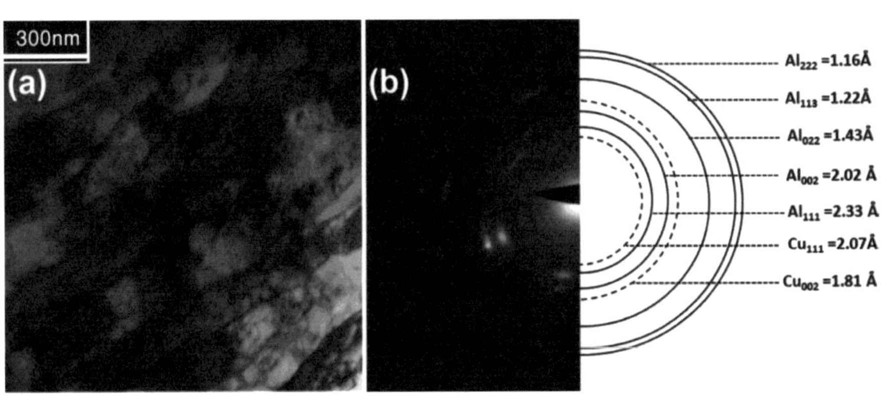

Figure 1.
(a) TEM image and (b) SAED of Al 2024 alloy under ECAP process [3].

of two alloys demonstrate streaks effects on spots pattern owing to highly pile-up of dislocations. Consequently, incomplete sectors of the rings in patterns were formed owing to preferred orientation in the crystal structure. Spots in the SAED pattern vary from a common shape to points with short and long tails. TEM images of nanostructured alloys and corresponding SAEDPs of the two samples demonstrated in **Figure 2**.

2.3 Kikuchi patterns

Kikuchi pattern consists of spots and paired parallel dark and bright lines. The spacing between a paired parallel dark and bright-line and the angle between Kikuchi lines in the pattern specify crystal structure characteristics such as the set of reflecting planes and distance of paired lines. By enhancing the thickness of the sample, the spots become more visible and the Kikuchi lines become less visible. Kikuchi patterns usually include spots and Kikuchi lines. Generally, Kikuchi patterns represent much more data rather than the spot patterns about crystal structure. The appearance of obvious Kikuchi pattern expresses a symptom of the ideal crystal. **Figure 3** displays the Kikuchi pattern of γFe with structural specifications [3].

Figure 4(a) illustrates TEM image of nanostructured Al 7075 alloy under thermomechanical processing. Tilting processing of the sample in TEM was applied to investigate the crystal lattice defects and orientation relationships. The SAED pattern includes the Kikuchi lines and spots of Al 7075 alloy was shown in **Figure 4(b)**.

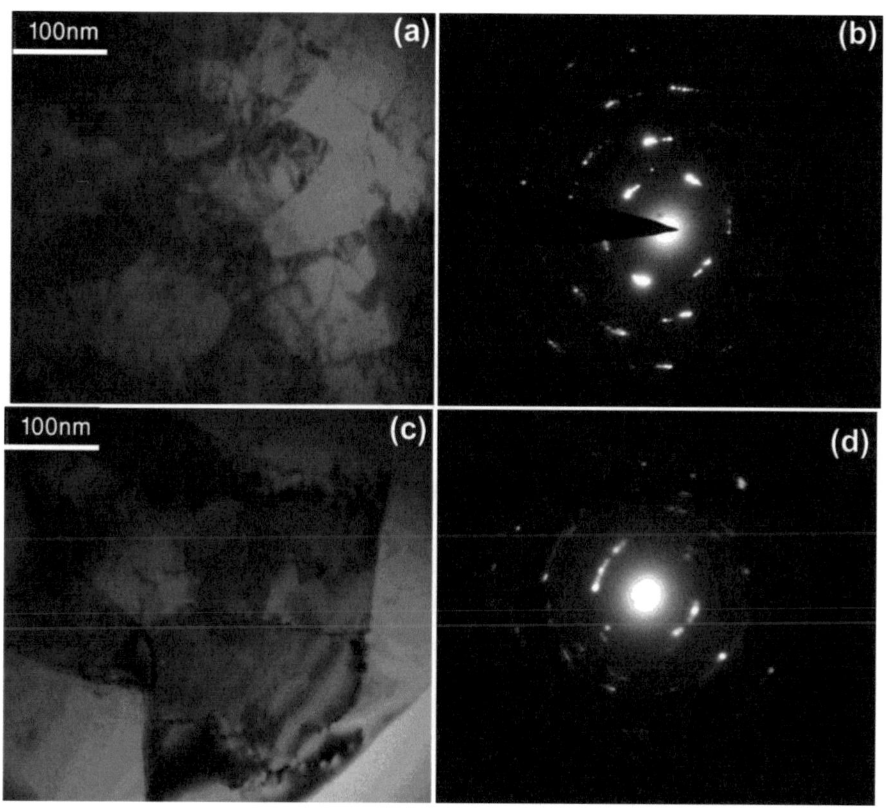

Figure 2.
(a) TEM images and (b) corresponding SAEDP of Al 5083 alloy under ECAP process [3], (c) TEM images and (d) corresponding SAEDP of Al 1050 alloy under CCR process [4, 5].

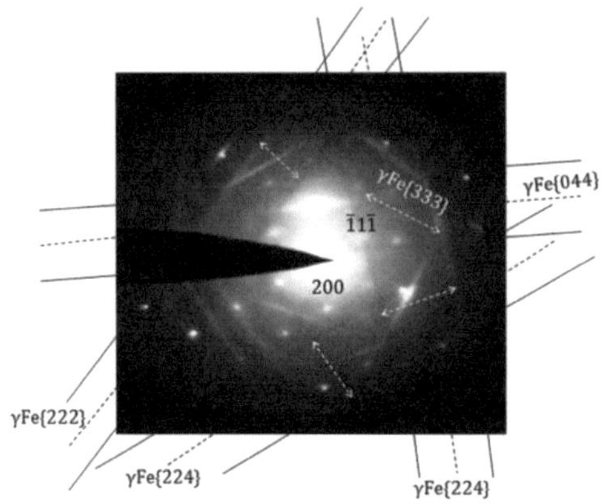

Figure 3.
Kikuchi pattern of γFe [3].

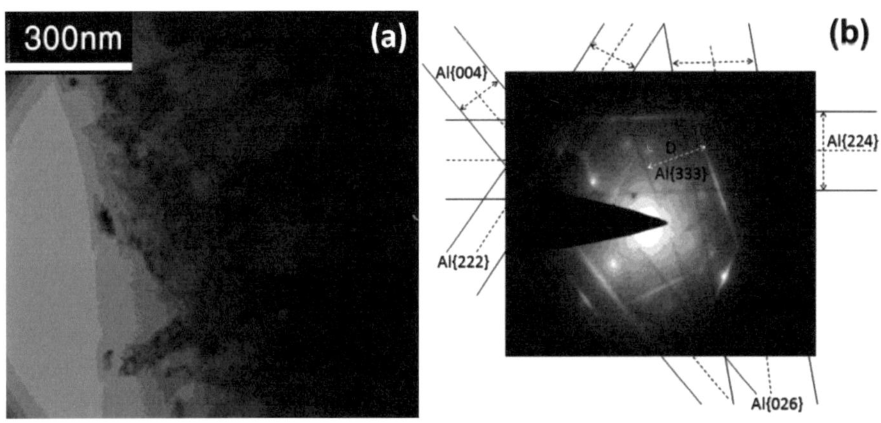

Figure 4.
(a) TEM image and (b) SAED pattern of nanostructured Al 7075 alloy [3].

The Kikuchi lines in SAED pattern were indexed the paired Kikuchi lines. Besides, crystal structure characteristics such as interplanar spacing and planes can be specified by measuring the distance between the paired Kikuchi lines [3].

2.4 Twinning

The twinning phenomenon has been known as the crystal structure defect and emerged principally as two parallel planes in TEM observations. In the twinning phenomenon, extra spots were produced surrounding the original spots of crystal structure in the pattern owing to the difference in orientation of twinning planes from the local crystal structure. TEM image and corresponding SAED pattern of twinning planes in *γFe* matrix with a f.c.c crystal lattice were indexed and shown in **Figure 5**, respectively. The main spots corresponding matrix pattern and twin spots with zone axis z = $[\bar{1}23]$ were mirror reflections across the $(1\bar{1}1)$ [3, 6].

2.5 Core-shell nanoparticles

Investigation results of the fabrication of Au/Al_2O_3 core-shell nanoparticles were demonstrated by one-step, and a very fast method with continuous-wave fiber laser ablation on an Aluminum (Al) plate coated gold (Au) nanolayer film immersed into ethanol. The core and shell of these nanoparticles contained complete crystalline structures. **Figure 6(a)** and **(b)** illustrates TEM images of Au/Al_2O_3 core-shell nanoparticles fabricated by the ablation of Al surface coated with Au a nanolayer. The shape of core-shell nanoparticles was approximately spherical and the single core-shell nanoparticle contains Au (core) and Al_2O_3 (shell), as shown in **Figure 6(b)**. The core and shell of the nanoparticles sometimes seemed entirely regular or non-regular. The core and shell of nanoparticles are determinable by the difference in contrast in TEM images. The structure of single core-shell nanoparticle includes completely crystalline gold (dark zone) coated by Al_2O_3 layer with crystalline structure (gray zone), as shown in **Figure 6(a)** and **(b)**. The Al_2O_3 shell of core-shell nanoparticle progressively has changed to the steady Al/Al_2O_3 compound by overtime. Phase characterization and crystalline planes of Au, Al, and Al_2O_3 were carried out by the use of the SAED technique, as shown in **Figure 6(c)** and **(d)** [7].

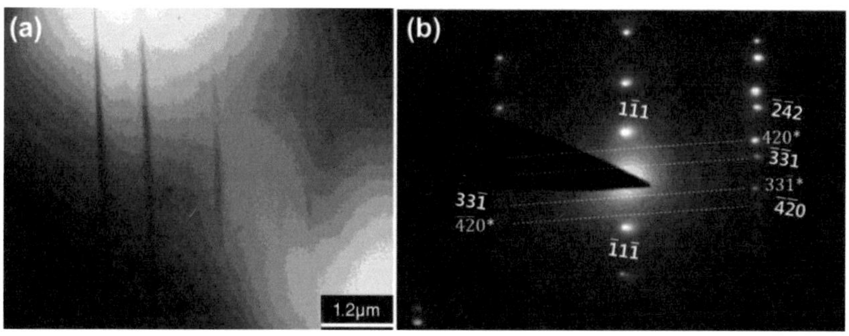

Figure 5.
(a) TEM image of twinning planes and corresponding SAED pattern of γFe *with f.c.c crystal lattice [3, 6].*

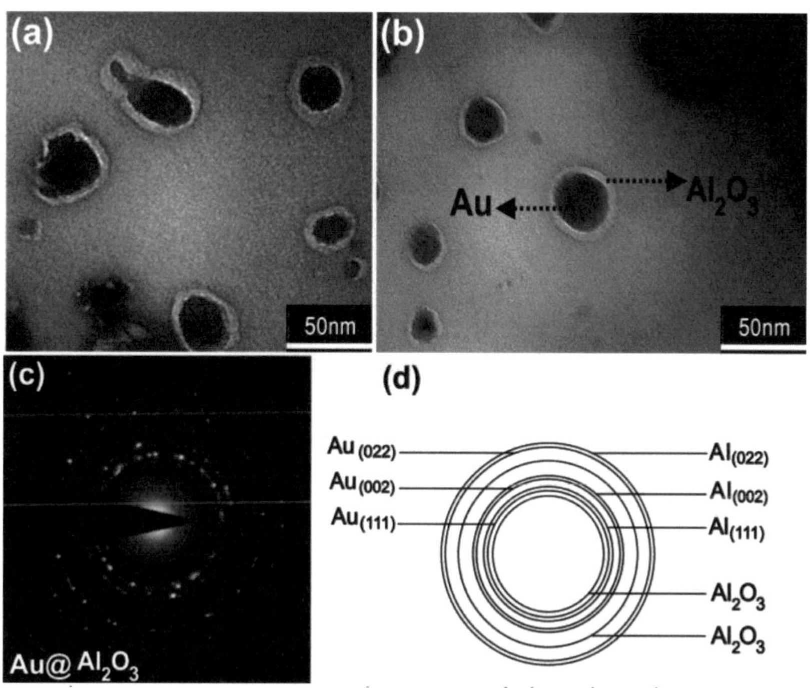

Figure 6.
(a, b) TEM images, (c) corresponding SAED and (d) analysis of ring pattern to identify phase and crystal planes of Au/Al_2O_3 core-shell nanoparticles [7].

By using of the EDP of individual nanoparticle, many of the crystalline structural properties can be measured such as interplanar spacing and the angles of the planes. TEM image of an individual Au/Al_2O_3 core-shell nanoparticle (**Figure 7(a)**) and the EDP of Au as a shell (**Figure 7(b)**) were demonstrated. In addition, the zone axis and planes of the shell were determined in accordance with standard patterns in **Figure 5(c)** and **(d)**. The spot and ring EDPs of core-shell nanoparticles corroborated of their crystalline specifications. By using of follow equation, lattice parameter of the shell was obtained.

$$\frac{1}{d^2} = \frac{h^2 + k^2 + l^2}{a^2} \tag{1}$$

where a is lattice parameter, d is interplanar spacing, and hkl are Miller indices. Lattice parameter of the core was calculated by using of Eq. (1) about a = 4.0502 Å which was near to the gold lattice parameter. Therefore, surely the core of these core-shell nanoparticles was gold [7].

2.6 Carbon nanotubes

Different types of carbon nanotubes (CNTs) have been one of the high potential materials for application in nanotechnology. In this work, the basic features of multiwall carbon nanotubes (MWCNTs) direct production on metal catalyst particles via thermal chemical vapor deposition (T.CVD) with existence of zirconium hydride (ZrH_2) and Al nanoparticles were investigated. **Figure 8(a)** and **(b)** illustrate TEM image of MWCNTs that the catalyst nanoparticles are various parts of nanotubes. The average size of the catalyst nanoparticles was about 20 nm. Thus, the considerable effect of ZrH_2 particles on the decreasing of the metal particles was clearly specified. Moreover, **Figure 8(c)** and **(d)** indicate the physical properties of MWCNTs such as the well-graphitized walls and open caps of them, respectively [8].

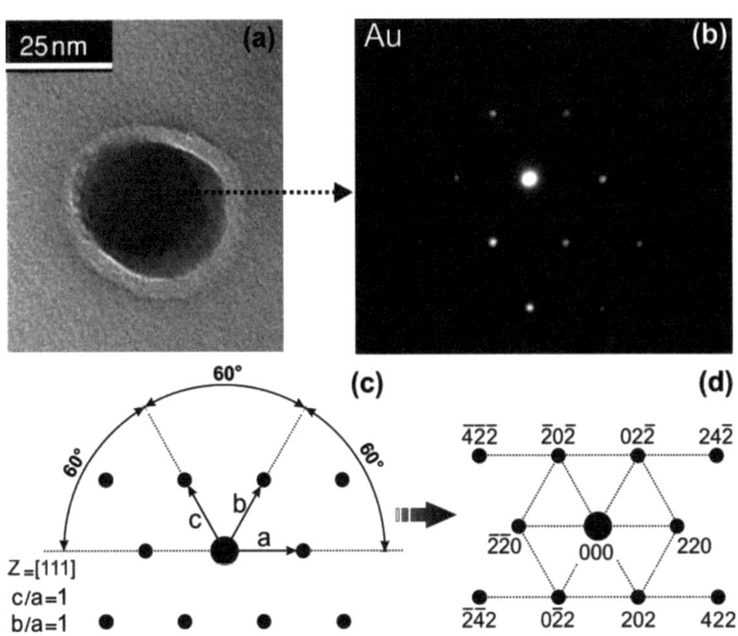

Figure 7.
(a) TEM image of Au/Al_2O_3 core-shell nanoparticles, (b) corresponding SAED of single nanoparticle core, (c, d) analysis of spot pattern to determined phase and crystal planes [7].

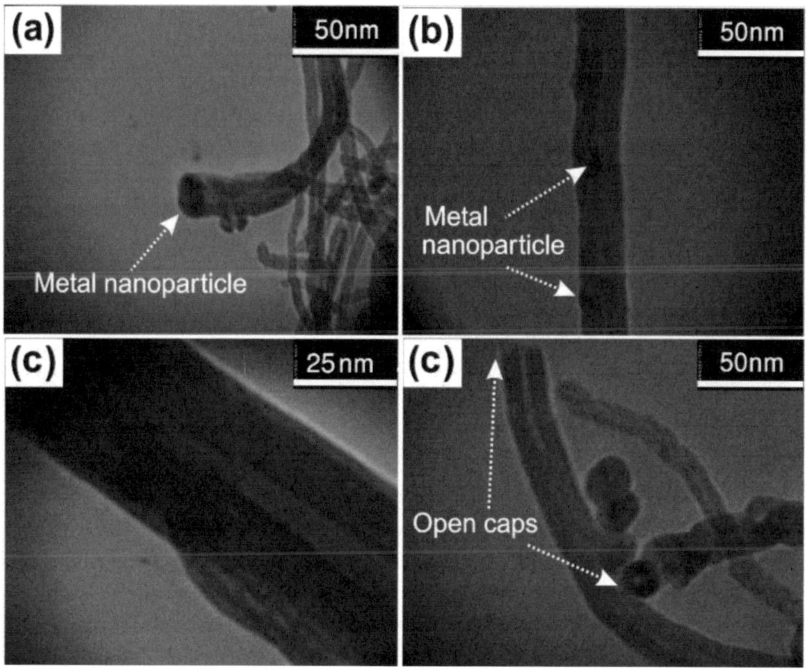

Figure 8.
TEM images of MWCNTs (a), (b) with catalyst nanoparticles, (c) and (d) with physical properties [8].

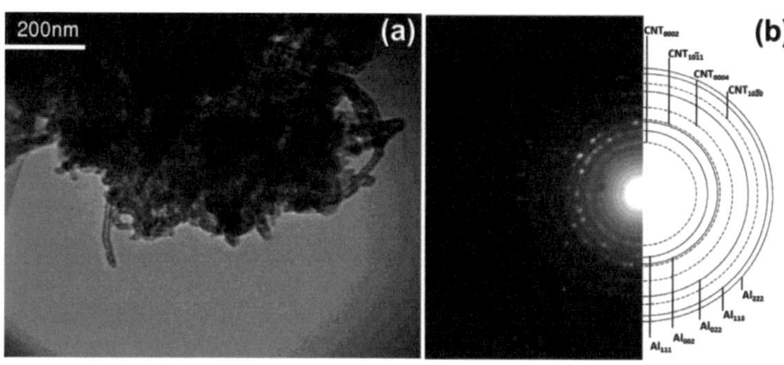

Figure 9.
(a) TEM image and (b) the SAED pattern of MWCNT/Al composite [9].

Compound of carbon nanotubes (CNTs) and metal nanoparticles has displayed considerable features. The major problem in CNTs-reinforced compound has been the regular dispersion of CNTs in nanoparticles. Ball milling technique as an effective method has been progressed for dispersion of the CNTs in micro and nanoparticles. For instance, 20 wt. % MWCNT+Al microparticle was compounded by ball milling method and studied using TEM and EDP technique as well. **Figure 9(a)** and **(b)** demonstrate TEM image and corresponding EDP of MWCNT/Al nanoparticles composite, respectively. Type of phases and crystal planes of MWCNTs and Al nanoparticles were identified, as shown in **Figure 9(b)** [9].

3. Transmission electron microscopy in the study of Al/BN Nanocomposite

The fabrication of bulk Al-BN nanocomposites have strongly been considered due to several advantages such as their light weight, superior mechanical properties even at low contents of BN as a reinforcement, and good thermal stability at elevated temperatures. These advantages are definitely related to the special characteristic of BN. It is well recognized that BN decomposed and dissolved in the Al matrix during high energy ball milling. Then, the possible AlN and AlB_2 as the in-situ phases are generated during a heating state in the Al matrix [10–17]. Due to presence of in-situ particles in matrix, the microstructural observation and mechanical behavior of the in-situ Al/BN nanocomposite should be investigated by high resolution transmission electron microscopy.

3.1 Phase evaluation

In the recent study, Al-1,2 and 4 wt. % BN bulk nanocomposite was fabricated by planetary ball milling of the composite powders and a post-process of hot extrusion at 580°C. As can be seen in **Figure 10**, the rod-shape phase was detected in TEM observation. Although AlN and AlB_2 as the in-situ phases were expected to create in microstructure, the accurate investigation using STEM has shown different results. As can be seen in **Figure 11**, the nature of the rod-shape phase was recognized by high-resolution STEM micrographs. The plane of atoms is clearly visible in the micrograph. The inter-planar spacing as the finger effect can lead to phase characterization. This parameter for the rod-shaped phases is measured to be about 0.55 nm, which is corresponding to the (006) planes of unwanted in-situ Al_4C_3 phase, respectively.

Figure 12 shows the bright-field TEM image of Al_4C_3 phase with the corresponding result of the EDS line scan the matrix and in-situ phase in the Al-2 wt. % BN nanocomposite. As can be, the high amount of carbon in analysis confirms the presence of Al_4C_3. In addition, the decreasing blue line and increasing the gray line in scanning through the in-situ phase corresponded to the contents of Al and carbon elements, respectively which were proved the formation of Al_4C_3.

The EDS analysis as a valuable technique in TEM is used to characterize Al_4C_3 in-situ phases in the microstructure of the Al-2 wt. % BN nanocomposite (**Figure 13**). Although the EDS technique does not accurately determine the composition, the results show the increasing the carbon content in the in-situ phase against the matrix in composition. Stearic acid as the process control agent (PCA) was recognized as the source of carbon in the nanocomposite. In fact, the PCA dissolved in the matrix during the planetary ball milling and Al_4C_3 was formed after hot extrusion as following reaction [18]:

$$4Al + 3C = Al_4C_3 \quad (2)$$

The SAED pattern of nanocomposite was shown in **Figure 14**. The continuous ring shape in the SAED pattern confirms that the nano crystallite structure was developed in the nanocomposite. In addition, the Al_4C_3 (012) and Al_4C_3 (0015) rings were observed in this pattern. Although the characterization of phases with the light elements such as nitrogen is troublous, the presence of AlN phase was shown in **Figure 14**. The diffraction reflections of AlN (100) in the SAED pattern indicate its in-situ formation through the extrusion process by the following equation:

$$Al + N = AlN \quad (3)$$

AlN phase as spherical shape and nanometric size was found in microstructure observation using STEM and the EDS analysis (**Figure 15**). As can be seen, increasing the nitrogen content compared to the matrix confirms the creation of

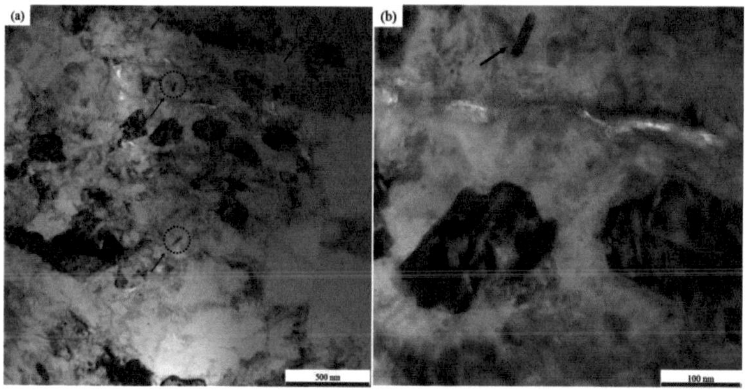

Figure 10.
TEM micrograph of Al-1 wt. % BN representing (a) the uniform distribution of a rod-shape in-situ phase and (b) higher magnification.

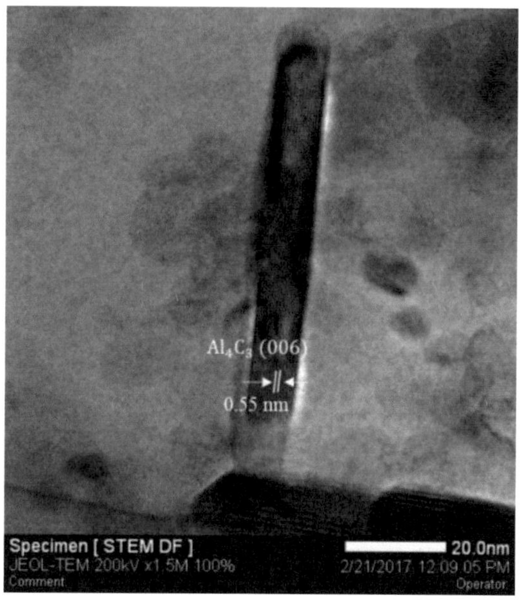

Figure 11.
High resolution STEM image of the in-situ Al_4C_3 phase with orientation of (006).

AlN in the matrix. Furthermore, the green line in the result of the EDS line scanning shows the higher nitrogen content of the spherical shape phase. According to the composition table in point (1), the high fraction of Nitrogen and all the Boron content remained as a solid solution in the matrix. These elements appeared as a solid solution during high energy ball milling in the Al matrix. Therefore, AlN phase was created with the reaction of Al and N in the hot extrusion process. On the other hand, AlB_2 which can be formed from the reaction of Al and B was not found.

Figure 16 shows the fraction of Al_2O_3 phase in the microstructure. As can be seen, the high content of oxygen compared to the matrix is clearly observed as a yellow line in the result of line scanning and the EDS analysis. Therefore, due to the limitation of XRD to detect the in-situ phase which has less than 5 vol. % in the matrix, the phase evaluation of this nanocomposite with the nano- and low content of phases is impossible using XRD.

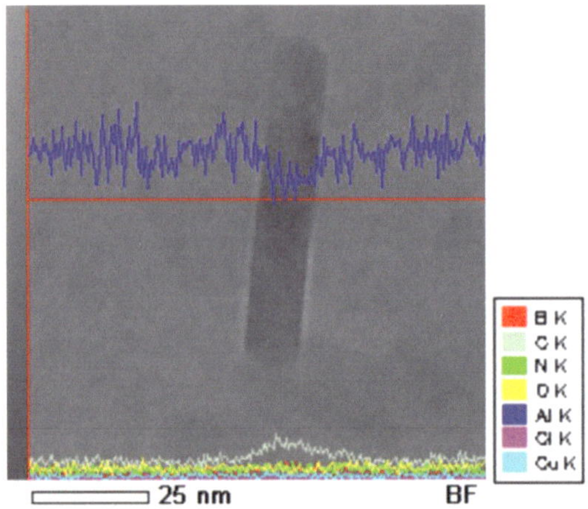

Figure 12.
Bright field TEM image of Al4C3 phase with the corresponding result of the EDS line scan.

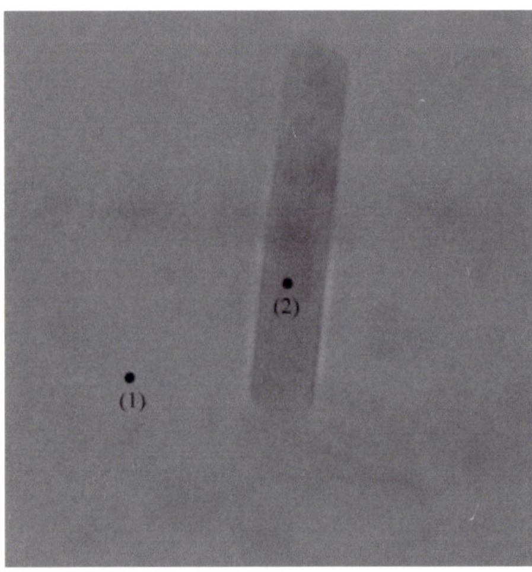

Elements	Atomic Percent (point 1)	Atomic Percent (point 2)
Al	77.07	48.77
C	14.73	43.4
N	5.95	5.88
O	2.25	1.95
B	-	-

Figure 13.
Bright field TEM images and the corresponding results of EDS analysis of the matrix and in-situ Al4C3 phases in Al-4 wt. % BN nanocomposite.

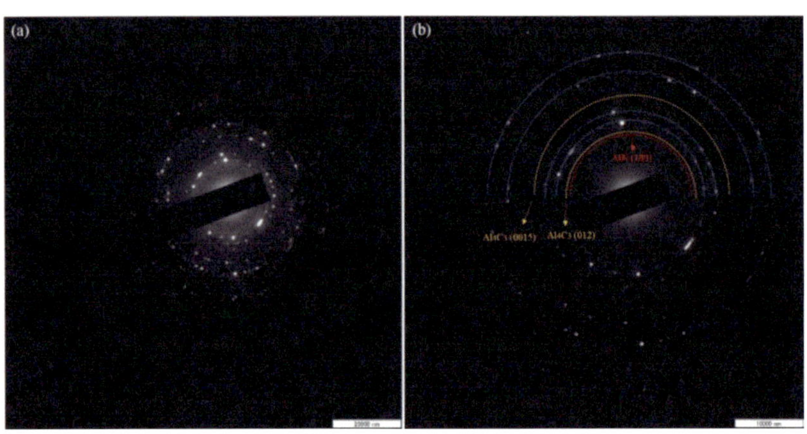

Figure 14.
(a) SAD pattern of Al-4 wt. % BN nanocomposite (b) higher magnification of SAD pattern of Al phase (blue color), the in-situ Al4C3 (yellow color) and AlN (red color) phases.

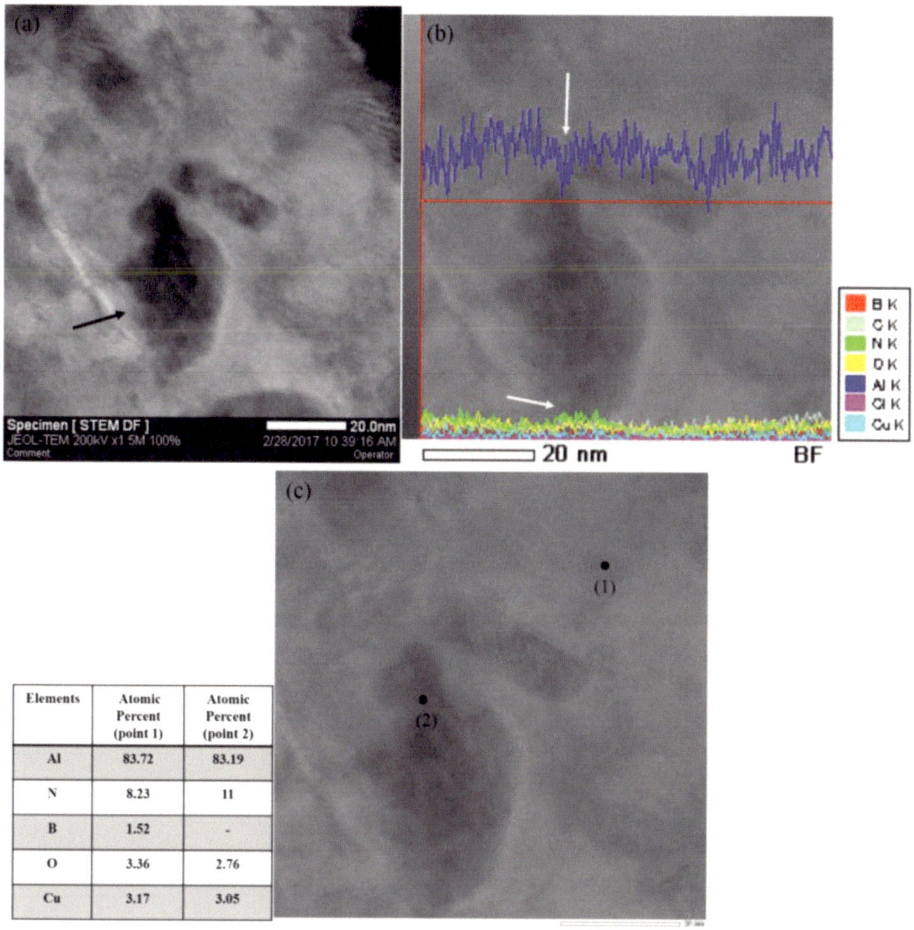

Elements	Atomic Percent (point 1)	Atomic Percent (point 2)
Al	83.72	83.19
N	8.23	11
B	1.52	-
O	3.36	2.76
Cu	3.17	3.05

Figure 15.
In-situ AlN phases in Al-4 wt. % BN nanocomposite (a) high resolution STEM image, (b) the EDS line scanning analysis, and (c) composition of the Al matrix and AlN phases.

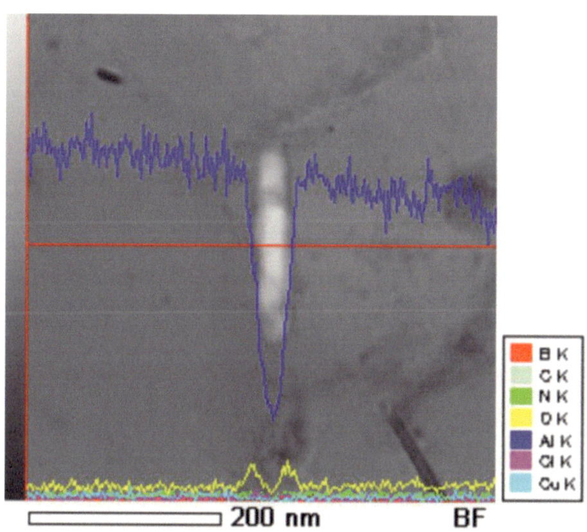

Figure 16.
The EDS line scanning analysis of Al_2O_3 phases in Al-1 wt. % BN nanocomposite.

3.2 Dislocation

Dislocation recognized as a linear defect is the justifier deformation mechanism. Density dislocation is a key factor is Calculated or estimated with a different method. Absolutely, the measuring corresponding to the observation of dislocations has more accurate than estimation with equations. In the mentioned nanocomposite, the TEM observation in **Figure 17** shows the cell wall dislocation and pile-up dislocation behind the grain boundary. As can be seen, the grain boundary in grain 1 acts as an obstacle against the dislocation movement. It led to pile up dislocation and stored strain in grain 1, while relaxation happened in the other adjacent grains with low dislocation density.

According to the following equation and TEM observations, the density dislocation (ρ) is calculated as a function of average separation between dislocations ($\bar{l}$) [19].

$$\bar{l} = 1/\rho \wedge 0.5 \tag{4}$$

Based on TEM observations and mentioned above equation, the dislocation density in the cell walls, accumulated behind the grain boundaries and in the grain interior are measured as $1.8 \times 10^{16}\ m^{-2}$, $7 \times 10^{14}\ m^{-2}$, and $1.5 \times 10^{14}\ m^{-2}$, respectively.

Figure 18 shows the incidence of extended dislocation was intersected on two slip planes. In this phenomenon, the new partial dislocation was created by intersecting of two leading partial dislocation, which is known as a stair-rod dislocation. In this dislocation, three dislocations were connected by wedge shape stacking fault and were immobile which is named Lomer-Cottrell lock. This is a barrier that pins the dislocation and led to work hardening in FCC materials [20].

Orowan dislocation loops in **Figure 19** indicate that dispersed phases such as Aluminum oxide particles in this nanocomposite act as the obstacle in dislocation movement.

3.3 Microstructural observations

Figure 20 displays the microstructure of nanocomposite after the hot extrusion process. As can be seen, the low angle boundary with the thick wall with an average thickness of 20 nm is surrounded by high angle boundaries. The cell

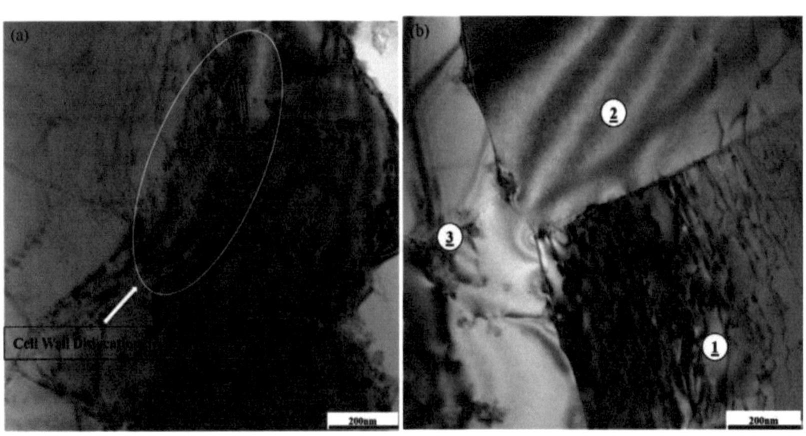

Figure 17.
TEM observation of dislocation in Al-4 wt. % BN nanocomposite (a) dislocation in cell wall and pile up dislocations behind the cell wall (b) accumulated dislocation near a grain boundary.

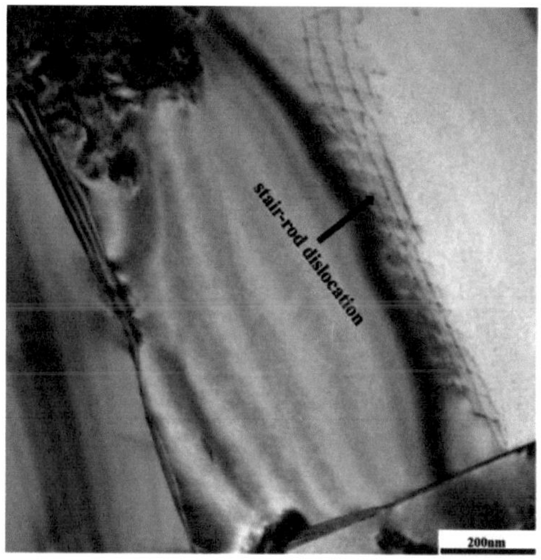

Figure 18.
Image of in Al-4 wt. % BN nanocomposite.

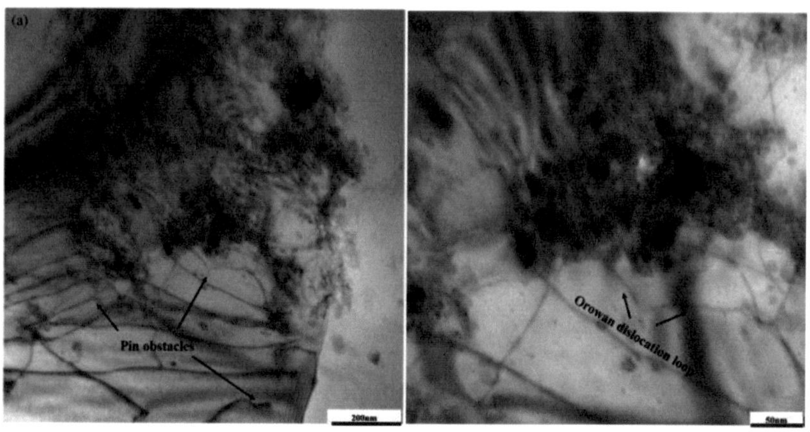

Figure 19.
Orowan dislocation loop (a) pin obstacles (b) higher magnification of Orowan dislocation loop.

structure happens in hot deformation of high SFE materials with converting three-dimensional dislocation tangles to two-dimensional dislocations with the more than 10° crystallographic misorientation. Then, the dislocation substructure is obtained from cell structure when the misorientation of cells exceeds to 2°. The dislocation substructures are converted to high angle sub boundaries which are surrounded by high angle boundaries [20].

Figure 21 shows the microstructure of Al-1 wt. % BN nanocomposite from the extrusion direction and cross-section of the extruded rod. As can be seen, the matrix consists of recrystallized nano/ultrafine grains with high angle boundaries and free dislocation density. It means the in-situ nanocomposite has thermal stability against the abnormal grain at high extrusion temperature.

Based on TEM observations and the high true strain value in hot extrusion process (2.3 mm/mm) which is higher than required critical strain for happening dynamic recrystallization in Al matrix (0.5 mm/mm), the recrystallization is the dominant mechanism in hot extrusion and high fraction of boundaries are formed as the high angle in microstructure [21]. On the other hand, the low angle boundaries and the grains with the high dislocation density were observed in microstructure

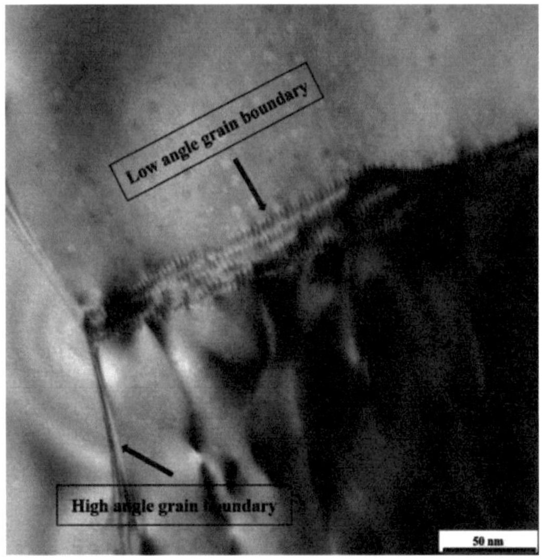

Figure 20.
*Image of low angle grain boundary surrounded by high angle grain boundaries (Higher magnification views from **Figure 19(b)**).*

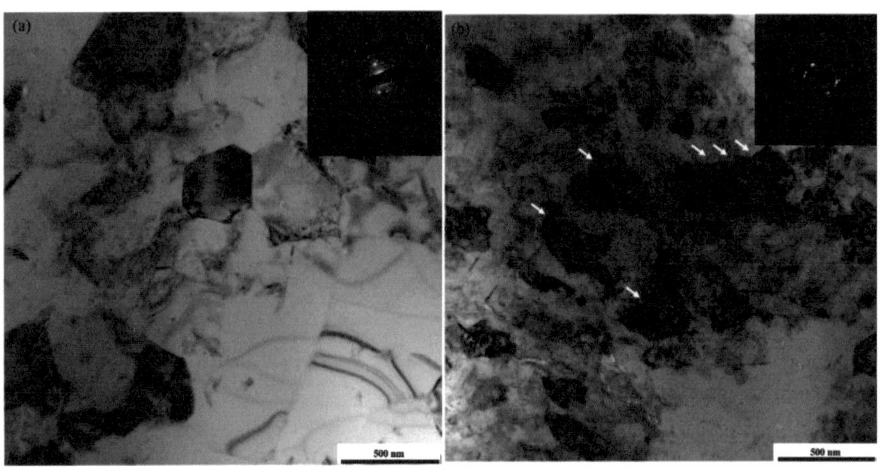

Figure 21.
Microstructure of Al-1 wt. % BN nanocomposite from (a) cross section of extruded rod (b) extrusion direction. Arrows show the scalloped boundaries.

(**Figure 22**). As can be seen, it seems that the rotation and coalescence is the main mechanism to decrease the low angle boundary. Due to high extrusion temperature, high stacking fault energy of Al matrix and high straining, the dynamic recovery occurred in the initial stage of deformation and the dislocations rearrangement themselves as low angle boundaries and subgrains is formed.

The geometrically necessary dislocation in metal matrix composite is high and led to increase in the kinetics of recrystallization [22–24]. However, the identification of dynamic recrystallization mechanisms in materials is difficult. The continuous dynamic recrystallization (CDRX) and the geometrical dynamic recrystallization (GDRX) can also occur simultaneously. During deformation of high stacking fault energy materials in elevated temperature, dynamic recovery prevents the accumulation of dislocation and the occurrence of discontinuous dynamic recrystallization (DDRX) will sustain. In CDRX, the subgrains are developed and

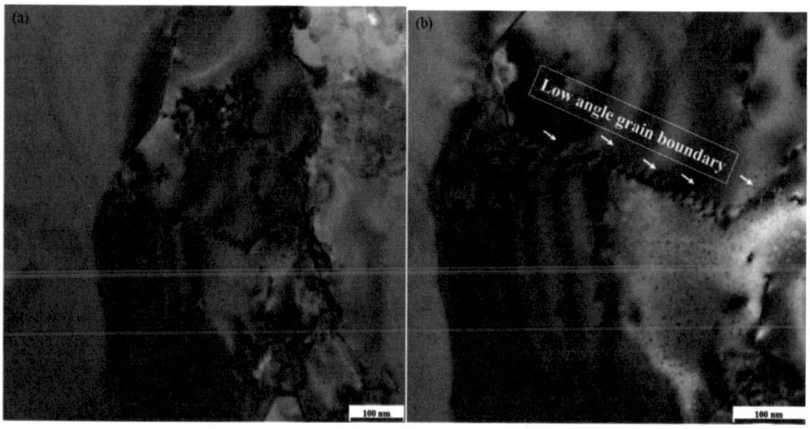

Figure 22.
(a) Microstructure of Al-4 wt. % BN nanocomposite from and (b) higher magnification.

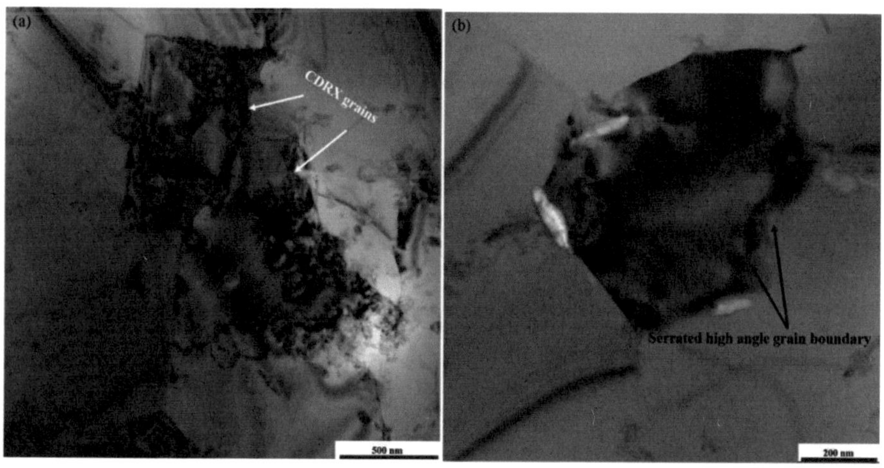

Figure 23.
Bright field TEM images of recrystallized grains in Al-4 wt. % BN (a) CDRX grains (b) GDRX grains.

then with increasing the misorientation of subgrain boundaries, the high angle grain boundaries will appear. On the other hand, in pure, solute drag and particle-containing alloy systems with high stacking fault energy which are deformed to large strain at high temperature, the GDRX will take place. In GDRX, the serrations are developed on the high angle boundaries and the scalloped boundaries are formed. In large straining, the grain elongation and thinning occurred. Then, the impingement of serrated boundaries led to equiaxed grains formation [22–26]. **Figure 23(a)** shows the ultrafine grains were recrystallized by CDRX. As can be seen in **Figures 21** and **23(b)**, the serrated high angle boundaries were observed. According to TEM results, the CDRX and GDRX are recognized as the dominant mechanisms in high angle grain boundary formation.

4. Conclusions

In summary, TEM/STEM with an energy-dispersive X-ray spectrometry corresponding to selected area diffraction patterns (SADP) is a powerful and accurate instrument to analysis and results from interpretation of nanomaterial lattice structure. By using its technique, most of the nanomaterial structural properties can be characterized such as orientation relationship determination, phase identification, twinning, second phases, crystallographic information, dislocation, preferred orientation and texture, extra spots and streaks. In addition, steps of structural changes (lattice parameter and interplanar spacing) and reducing of nanoparticles and nano-grains sizes of different materials can be investigated by taking the EDPs of nanoparticles synthesis or nanostructuring process steps.

Acknowledgements

The authors would like to thank the TEM laboratory in Advanced Materials Group of Materials Research School for their financial supports.

Author details

Mohammad Jafari Eskandari[1*], Reza Gostariani[2] and Mohsen Asadi Asadabad[2]

1 Department of Materials Engineering, Isfahan University of Technology, Isfahan, Iran

2 Nuclear Science and Technology Research Institute, Iran

*Address all correspondence to: m_jafari@ma.iut.ac.ir

References

[1] Sabirov I, Enikeev NA, Murashkin MY, Valiev RZ. Bulk Nanostructured Materials with Multifunctional Properties. Cham/Heidelberg/New York: Springer; 2015

[2] Murty BS, Shankar P, Raj B, Rath BB. Textbook of and Nanoscience Nanotechnology. India: Universities Press; 2013

[3] Asadi Asadabad M, Jafari Eskandari M. In: Janecek M, Kral R, editors. Electron Diffraction, Modern Electron Microscopy in Physical and Life Sciences. Rijeka: InTech; 2016. pp. 2-26

[4] Jafari Eskandari M, Asadi Asadabad M, Tafrishi R, Shafyei A. Evolution of nanostructure in Al 1050 sheet deformed by cryo-cross-rolling. JMEPEG. 2016;**25**:1643-1649

[5] Jafari Eskandari M, Asadi Asadabad M, Tafrishi R, Bagherzadeh M. The effect of cross-rolling process on nanostructure of Al 1050 alloy. Journal of Nanoanalysis. 2014;**1**(2):93-98

[6] Asadi AM, Jafari EM. Transmission electron microscopy as best technique for characterization in nanotechnology. Synthesis and Reactivity in Inorganic, Metal-Organic, and Nano-Metal Chemistry. 2015;**45**:323-326

[7] Jafari Eskandari M, Shafyei A, Karimzadeh F. One-step fabrication of Au@Al_2O_3 core-shell nanoparticles by continuous-wave fiber laser ablation of thin gold layer on aluminum surface: Structural and optical properties. Optics and Laser Technology. 2020;**126**:106066

[8] Hasanzadeh I, Jafari Eskandari M. Direct growth of multiwall carbon nanotube on metal catalyst by chemical vapor deposition: In situ nucleation. Surface and Coating Technology. 2020;**381**:125109

[9] Jafari Eskandari M, Asadi Asadabad M, Tafrishi R, Emamalizadeh M. Transmission electron microscopy characterization of diffraction nanotubes. Inorganic and Nano-Metal Chemistry. 2017;**47**(2):197-201

[10] Xia Z, Li Z, Lu C, Zhang B, Zhou Y. Structural evolution of Al/BN mixture during mechanical alloying. Journal of Alloys and Compounds. 2005;**399**:139-143

[11] Firestein KL, Steinman AE, Golovin IS, Cifre J, Obraztsova EA, Matveev AT, et al. Fabrication, characterization and mechanical properties of spark plasma sintered Al–BN nanoparticle composites. Materials Science and Engineering A. 2015;**642**:104-112

[12] Firestein K, Corthay S, Steinman A, Matveev A, Kovalskii A, Sukhorukova I, et al. High-strength aluminum-based composites reinforced with BN, AlB2 and AlN particles fabricated via reactive spark plasma sintering of Al-BN powder mixtures. Materials Science and Engineering A. 2017;**681**:1-9

[13] Gostariani R, Ebrahimi R, Asadi Asadabad M, Paydar MH. Mechanical properties of Al/BN nanocomposites fabricated by planetary ball milling and conventional hot extrusion. Acta Metallurgica Sinica (English Letters). 2017;**31**:245-253

[14] Gostariani R, Bagherpour E, Rifai M, Ebrahimi R, Miyamoto H. Fabrication of Al/AlN in-situ nanocomposite through planetary ball milling and hot extrusion of Al/BN: Microstructural evaluation and mechanical behavior. Journal of Alloys and Compounds. 2018;**768**:329-339

[15] Gostariani R, Asadi Asadabad M, Paydar MH, Ebrahimi R. Morphological and phase evaluation of Al/15wt.% BN nanocomposite synthesized by planetary ball mill and sintering. Advanced Powder Technology. 2017;**28**:2232-2238

[16] Lonardelli I, Zadra M, Ischia G, Barreiro JG, Bortolotti M, Molinari A. In situ synchrotron X-ray diffraction experiments on Ale15%BN mechanically alloyed powder: Observation of AlN nanoparticles precipitation and enhanced thermal stability of nanostructured Al matrix. Journal of Alloys and Compounds. 2009;**486**:653-659. DOI: 10.1016/j.jallcom.2009.07.024

[17] Gostariani R, Ebrahimi R, Asadi Asadabad M. The study of hot deformation behavior of mechanically milled and hot extruded Al-BN nanocomposite. Transactions of the Indian Institute of Metals. 2018;**71**:1127-1136. DOI: 10.1007/s12666-017-1248-x

[18] Suryanarayana C. Mechanical alloying and milling. Progress in Materials Science. 2001;**46**:1-184. DOI: 10.1016/S0079-6425(99)00010-9

[19] Soboyejo W. Mechanical Properties of Engineered Materials. New York: Marcel Dekker; 2002

[20] Zlateva G, Martinova Z. Microstructure of Metals and Alloys: An Atlas of Transmission Electron Microscopy Images. Boca Raton: CRC Press; 2008

[21] Lin Y, Xu B, Feng Y, Lavernia EJ. Stress-induced grain growth during hightemperature deformation of nanostructured Al containing nanoscale oxide particles. Journal of Alloys and Compounds. 2014;**596**:79-85. DOI: 10.1016/j.jallcom.2014.01.189

[22] Humphreys FJ, Hatherly M. Recrystallization and Related Annealing Phenomena. Oxford: Elsevier; 2004

[23] Singh D, Yun Y, Wan W, Grushko B, Zou X, Hovmöller S. A complex pseudo-decagonal quasicrystal approximant, Al37(Co, Ni)15.5, solved by rotation electron diffraction. Journal of Applied Crystallography. 2014;**47**:215. DOI: 10.1107/S1600576713029294

[24] Singh D, Yun Y, Wan W, Grushko B, Zou X, Hovmöller S. Hovmöller, structure determination of a pseudo-decagonal quasicrystal approximant by the strong reflections approach and rotation electron diffraction. Journal of Applied Crystallography. 2016;**49**:433. DOI: 10.1107/S1600576716000042

[25] Singh D, Shahi RR, Yadav TP, Mandal RK, Tiwari RS, Srivastava ON. Hydrogenation of (Zr69. 5Al7.5Cu12Ni11) 100-xTix quasicrystalline alloys and its effect on their structural and microhardness behavior. Journal of Non-Crystalline Solids. 2013;**380**:11-16. DOI: 10.1016/j.jnoncrysol.2013.08.024

[26] Singh D, Yadav TP, Tiwari RS, Srivastava ON, Singh M, Mandal RK. Nanoindentation studies of metallic glasses and nanoquasicrystal–glass composites in Zr–Al (Ga)–Cu–Ni alloys. International Journal of Nanoscience. 2011;**10**:929-933. DOI: 10.1142/S0219581X11008769

Chapter 3

Empires: The Nonlocal Properties of Quasicrystals

Fang Fang, Sinziana Paduroiu, Dugan Hammock and Klee Irwin

Abstract

In quasicrystals, any given local patch—called an emperor—forces at all distances the existence of accompanying tiles—called the empire—revealing thus their inherent nonlocality. In this chapter, we review and compare the methods currently used for generating the empires, with a focus on the cut-and-project method, which can be generalized to calculate empires for any quasicrystals that are projections of cubic lattices. Projections of non-cubic lattices are more restrictive and some modifications to the cut-and-project method must be made in order to correctly compute the tilings and their empires. Interactions between empires have been modeled in a game-of-life approach governed by nonlocal rules and will be discussed in 2D and 3D quasicrystals. These nonlocal properties and the consequent dynamical evolution have many applications in quasicrystals research, and we will explore the connections with current material science experimental research.

Keywords: quasicrystals, empire, nonlocality, Penrose tiling, cut-and-project

1. Introduction

Quasicrystals are objects with aperiodic order and no translational symmetry. These "peculiar" objects, deemed in and out of existence by theoretical considerations, have been discovered in 1982 by Shechtman [1], in agreement with previous predictions [2, 3]. Shechtman's discovery was honored with the award of the Nobel Prize in Chemistry in 2011. Believed to be rare in nature initially, up to this date there have been found roughly one hundred quasicrystal phases that exhibit diffraction patterns showing quasiperiodic structures in metallic systems [4, 5]. Quasicrystals were first constructed as aperiodic tilings defined by an initial set of prototiles and their matching rules; constructing these tilings meant aggregating tiles onto an initial patch so as to fill space, ideally without gaps or defects. Likewise, quasicrystals in physical materials are formed by atoms accumulating to one another according to the geometry of their chemical bonding. Curiously, the localized growth patterns would give rise to structures which exhibit long-range and nonlocal order, and mathematical constructions were later discovered for creating geometrically perfect, infinite quasiperiodic tilings of space. In material science, new electron crystallography methods and techniques have been developed to study the structure and geometrical patterns of quasicrystal approximants [6–8], revealing unique atom configurations of complicated quasicrystal approximant structures [7, 8].

Complex and varied in their structure, quasicrystals translate their intrinsic nonlocal properties into nonlocal dynamic patterns [9]. The empire problem is an investigation into the nonlocal patterns that are imposed within a quasicrystal by a finite patch, where just a few tiles can have a global influence in the tiling so as to force an infinite arrangement of other tiles throughout the quasicrystal [10]. Initial research into calculating empires—a term originally coined by Conway [11]—focused on the various manifestations of the Penrose tiling [12] such as the decorated kites-and-darts, where Ammann bars would indicate the forced tiles [10], and the multi-grid method, where algebraic constraints can be employed [13]. More recently, the cut-and-project technique [14, 15]—where the geometry of convex polytopes comes into play—has been implemented into a most efficient method of computing empires [16]. The cut-and-project method offers the most generality and has been used to calculate empires for the Penrose tiling and other quasicrystals that are projections of cubic lattices $\mathbb{Z}^n$ [16, 17]. Quasicrystals that are projections of non-cubic lattices (e.g., the Elser-Sloane tiling as a projection of the E_8 lattice to $\mathbb{E}^4$) are more restrictive and some modifications to the cut-and-project method have been made in order to correctly compute the tilings and their empires [18]. The cut-and-project formalism can also be altered to calculate the space of all tilings that are allowed by a given patch, wherein the set of forced tiles of the patch's empire is precisely the mutual intersection of all tilings which contain that patch [15].

The interest in the nonlocal nature of empires has led to further explorations into how multiple empires can interact within a given quasicrystal, where separate patches can impose geometric restrictions on each other no matter how far apart they may be located within the tiling. These interactions can be used to define rules (similar to cellular automata) to see what dynamics emerge from a game-of-life style evolution of the quasicrystal and for the first time such a simulation with nonlocal rules has been performed [17].

The empires can be used to recover information from the higher dimensional lattice from which the quasicrystal was projected, filtering out any defects in the quasicrystal and therefore providing an error self-correction tool for quasicrystal growth [16]. In terms of quasicrystal dynamics, the empires provide us with the opportunity of developing algorithms to study the behavior and interactions of quasicrystalline patches based on nonlocal rules—a very rich area of exploration.

In this chapter we review the nonlocal properties of quasicrystals and the studies done to generate and analyze the empires and we discuss some of the findings and their possible implications.

2. Empires in quasicrystals

Empires represent thus all the tiles forced into existence at all distances by a quasicrystal patch. When it comes to analyzing the forced tile distribution in a quasicrystal, we differentiate between the local and nonlocal configurations. The tiles surrounding the vertex, that is the tiles that share one vertex, form the vertex patch. The local empire is the union of forced tiles that are in the immediate vicinity of an emperor, be it a vertex or a patch, where there are no "free" tiles in between the emperor and the forced tiles. The forced tiles that are at a distance from the emperor form the nonlocal part of the empire.

2.1 Methods for generating empires

Several methods for generating the empires in quasicrystals have been discussed in [16]. The Fibonacci-Grid method employs the Penrose tiling decoration using Amman bars that form a grid of five sets of parallel lines [10, 19]. The grid constructed with the Amman bars is, in fact, a Fibonacci grid [20]. When Amman lines intersect in a certain configuration, a tile or a set of tiles are forced. The multigrid method [13] describes a tiling that can be constructed using a dual of a pentagrid, a superposition of five distinct families of hyperplanes in the case of Penrose tiling. It can be used for generating empires in cases where the dual grid for the quasicrystal has a simple representation, but it is not effective when the quasicrystal has a defect, as its dual is no longer a perfect multigrid.

The most efficient method of generating empires in quasicrystals is the cut-and-project method. While the cut-and-project method and the multigrid one are mathematically equivalent in their use of generating empires, the cut-and-project method provides us with the possibility of recovering the initial mother lattice, even for defected quasicrystals [16]. The method has also been applied to projections of non-cubic lattices, making it of general use [18].

When we project from a lattice $\Lambda \subset \mathbb{E}^N$ $\pi_{\|}, \pi_{\perp}$ onto orthogonal subspaces $\mathbb{E}_{\|}, \mathbb{E}_{\perp}$, the *cut-window*, which is a convex volume $\mathcal{W}$ in the perpendicular space $\mathbb{E}_{\perp}$, determines the points selected to have their projections included in the tiling of the subspace $\mathbb{E}_{\|}$, acting thus as an acceptance domain for the tiling. Once the tiling is generated, all the tiles in a given local patch can be traced back to the mother lattice, giving a restriction on the cut-window. The *possibility-space-window* represents the union of all cut-windows that satisfy the restriction—all the tiles in the possibility-space-window *can* legally coexist with the chosen patch. The *empire-window* represents the intersection of all the possible cut-windows—all the tiles inside the empire-window *must* coexist with the initial patch. This in turn acts as the cut-window for the forced tiles, the patch's empire.[1]

For cubic lattices ($\Lambda = \mathbb{Z}^N$) the cut-window is sufficient for determining the tiles which fill the tiling space $\mathbb{E}_{\perp}$, but it needs to be sub-divided into regions acting as acceptance domains for individual tiles, when projecting non-cubic lattices.

To review some of the results of applying the cut-and-project method to the calculation of empires, we will give several examples in 2D and 3D cases.

2.2 Empires in 2D

For the 2D case, we will consider the Penrose tiling, a non-periodic tiling, a quasicrystal configuration that can be generated using an aperiodic set of prototiles. The 2D Penrose tiling, when projected from the cubic lattice $\mathbb{Z}^5$, has eight vertex types: D, J, K, Q, S, S3, S4 and S5 [21, 22]. In **Figure 1**, we show the empires for three of the vertex types that do not present five-fold symmetry (D, K, S4) computed with the cut-and-project method along with their possibility space [16]. The S4 vertex has the densest empire, while the D vertex has no forced tiles. The density of the empire depends on the size of the empire window, individual for each vetex type. In **Figure 2**, we show the empires and the possibility space of the five-fold symmetry vertex types, S5 (the star) and S (the sun) [16]. The empires of all 8 vertex types are displayed in [16].

2.3 Empires in 3D

The cut-and-project method described above can be used also for computing the empires of a given patch in 3D, e.g., an Amman tiling defined by a projection of $\mathbb{Z}^6$ to $\mathbb{E}^3$ [18]. In **Figure 3**, we show three orientations of the empires of two of the vertex types for this projection and we can see they differ in structure, as well as in density. Two other vertex types together with their empires are shown in [18].

---

1 For a detailed description of this method and a comparison with the other methods, please refer to [16].

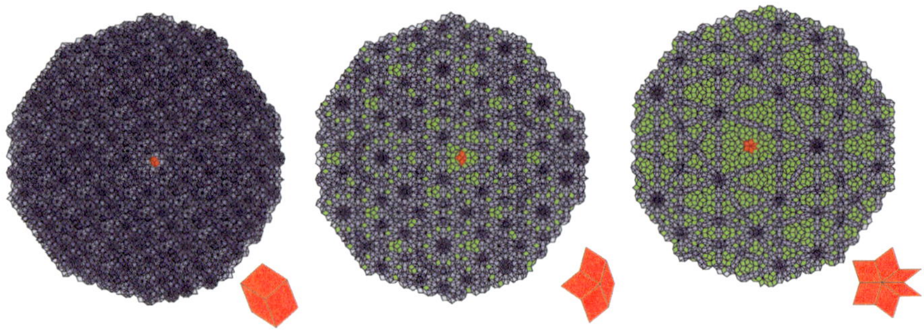

Figure 1.
Empire calculation of the vertex types D (left), K (middle) and S4 (right). The red tiles represent the vertex patch and the green tiles represent the empire. The enlarged vertex patch is shown in the right corner for each case. The pictures for all vertex types have been published in [16], Figures 13 and 14.

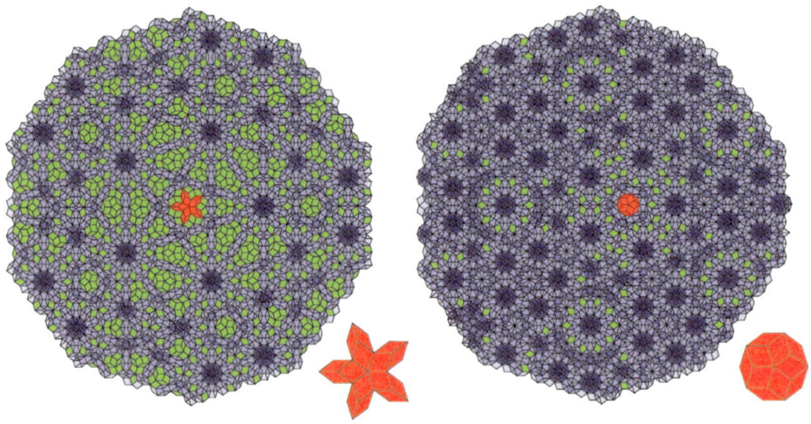

Figure 2.
Empire calculation for the vertex types S5 (left) and S (right). The red tiles represent the vertex patch and the green tiles represent the empire. The enlarged vertex patch is shown in the right corner for each case. The pictures for all vertex types have been published in [16], Figures 13 and 14.

Furthermore, the method can be used also for non-cubic lattices, but it requires some adjustment, as the proper selection of the tiles becomes more complex. It has been shown [18] that the cut-window must be sub-divided into regions, which act as acceptance domains for individual tiles and can be used to compute both the relative frequencies of the vertex's configuration and the empire of a given tile configuration. The authors have used the method to compute the frequencies and sectors for an icosahedral projection of the D_6 lattice to $\mathbb{E}^3$, which has 36 vertex configurations [23] and have compared their findings with previous results [23–25].

3. Quasicrystal dynamics

The inherent nonlocal properties of quasicrystals allow us to study different dynamical models of self-interaction and interactions between different vertex configurations in quasicrystals, using the empires. Several game-of-life [21] algorithms have been previously studied on Penrose tiling, but they have either considered a periodic grid [26] or they have considered only local rules [27, 28]. Recently, for the first time, a game-of-life scenario has been simulated using nonlocal rules on a two-dimensional qusicrystal, the Penrose tiling, in [17]. In this simulation, for the

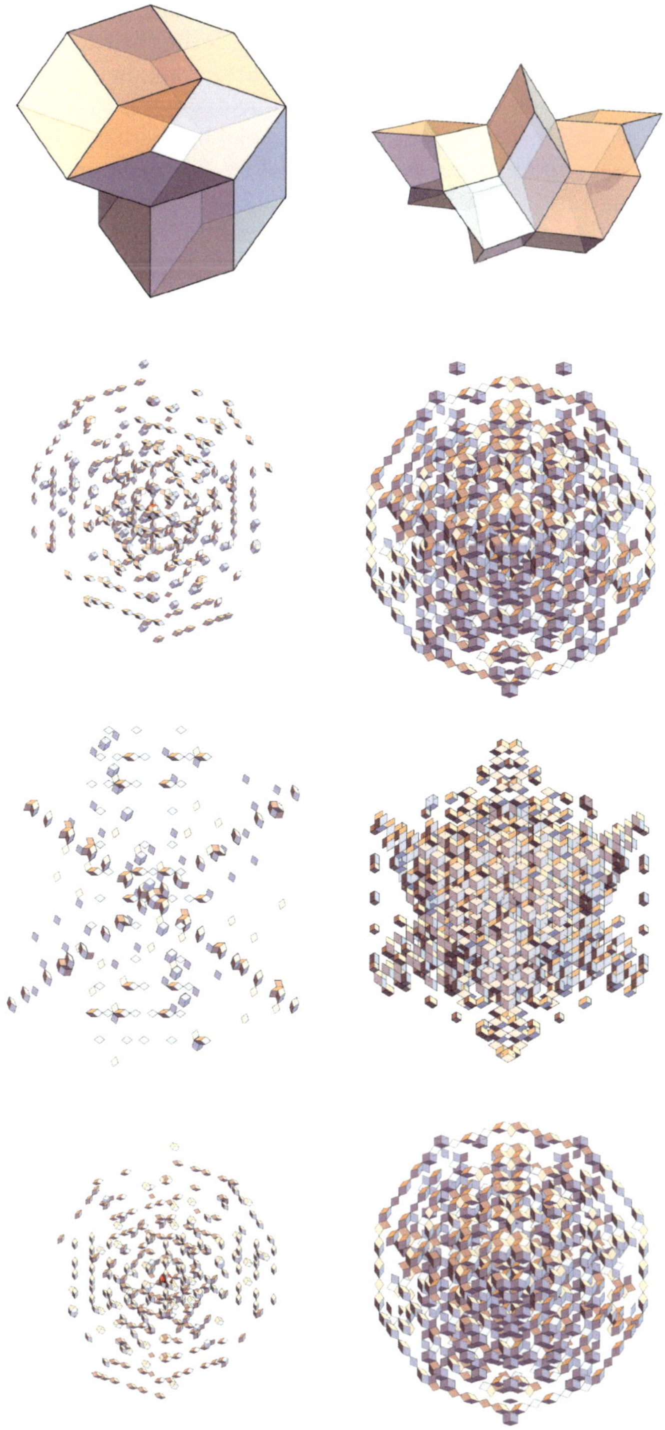

Figure 3.
Vertex configurations and their empires for the Ammann tiling as projected from $\mathbb{Z}^6$ to $\mathbb{E}^3$. The tiles of this quasicrystal are all rhombohedrons, and the vertex configurations are analogous to those of the Penrose tiling. The empires are shown in three orientations below the vertex configurations. Other two vertex configurations and their empires have been shown in [18] Figure 9.

K vertex type, the emperor and its local patch are treated as a quasiparticle, a glider. The empire acts as a field and the interaction between two quasiparticles is modeled as the interaction between empires.

Several rules have been employed to describe both the self-interaction and the two-particle interactions. Firstly, the neighbors where the vertex patch is allowed to move are constrained by the higher dimensional projection, being the closest neighbors of the same vertex type in the perpendicular space. This approach differs from previous studies that consider the nearest neighbors situated in the local 2D representation [27, 28]. In **Figure 4**, one can see the K vertex type and 2D representation of the nearest neighbors in the perpendicular space that we have considered [17]. The distribution of the neighbors is interesting, as it surrounds an S vertex patch (sun) on one side and an S5 vertex patch (star) on the other side. We will expand on this configuration in the next section.

Secondly, the vertex never stays in the same position for two consecutive frames, being thus forced to move to one of the allowed neighbors in its immediate vicinity. Depending on the intrinsic configuration of the vertex patch, some vertices allow more freedom of movement than others. For example, a vertex with a five-fold symmetry will tend to perform a "circular" motion around its axis of symmetry, a rotation, while a vertex without the five-fold symmetry, like the K vertex, will have the possibility to propagate forward, the translational movement being a sequence of rotations around different centers.

Moreover, the particle moves following the "least change" rule, which states that the particle should move to the position (or one of the positions) where the result of that movement implies that the number of tiles changed in the empire is minimum. In other words, the particle will follow the path that requires the least number of changes in the tiles in the empire, while not being allowed to stay in the same position for two consecutive frames. When there is more than one choice that obeys the aforementioned rules, a random-hinge variable is introduced such that the particle will chose one of the favored positions. Due to the syntactical freedom provided by this choice, the path of a particle, unless constrained otherwise, is impossible to predict with 100% accuracy.

For the case of two-particle interactions, one more arbitrary constraint is introduced, where the local patches of the two particles are not allowed to overlap. A detailed discussion of the algorithm and the simulation setup can be found in [17].[2]

Figure 4.
The K vertex type surrounded by the eight possible neighbors. The orange dots represent the position of the K vertices that surround the star and sun vertex patches.

---

[2] Movies from the simulations can be watched on https://www.youtube.com/playlist?list=PL-kqKe jCypNT990P0h2CFhrRCpaH9e858.

Besides the initial conditions, meaning the vertex type configuration and the initial position of the particles in the two-particle interaction case, the movement of the particles is influenced solely by their empires and their possibility space. When the empire changes, the possibility space changes as well, constraining the next move due to the new spatial configuration. The empire and the possibility space create a feedback loop of influence, which for an infinite quasicrystal propagates instantly at infinite distances.

One of the most interesting findings is that in the case of the two-particle interactions, the particles get locked in an oscillation type movement when they are in proximity. If we consider an analogy between the "least change" principle—the number of tiles that change between two steps is minimum—and a minimum energy principle, the system tends to reach a minimum energy state in oscillation. This is similar with the time crystal scenarios [29–32] where a system disturbed by a periodic signal reaches a quasistable state in which it oscillates at a period different from the period of the external kick. In this case, a quasiparticle will draw stability from its empire interaction with other quasiparticles' empires—a nonlocal induced stability.

4. Empires and higher dimensional representations

Quasicrystals projected from a higher dimensional lattice, $\mathbb{Z}^5$ for example, show several properties dictated from the representation in the high dimension, like the symmetry and the vertex and empire distribution. As discussed previously, for the K vertex type, the nearest neighbors considered in the game-of-life scenario are also dictated by the higher dimensional lattice from which the tiling is projected, being the closest neighbors in the perpendicular space [17].

In **Figure 4**, we have shown the K vertex type with its eight neighbors that surround a sun and a star configuration. **Figure 5** shows the sun vertex patch surrounded by the five orientations of the K vertex type. This is a more complex structure that has five-fold symmetry. When choosing only the sun configurations that are bordered by the K-type vertices, we observe that these configurations come from two different regions in the perpendicular space. In **Figure 6**, we show the 2D distribution of the K-type vertices, plotted in two different colors, corresponding to the two distinct regions in the perpendicular space from which the vertices are projected. The vertices form interesting patterns on the Penrose tiling. **Figure 7**

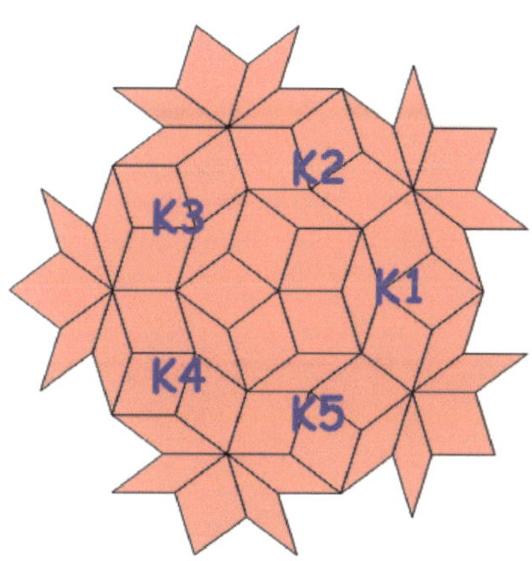

Figure 5.
The sun configuration surrounded by the five different orientations of the K-type vertex patch.

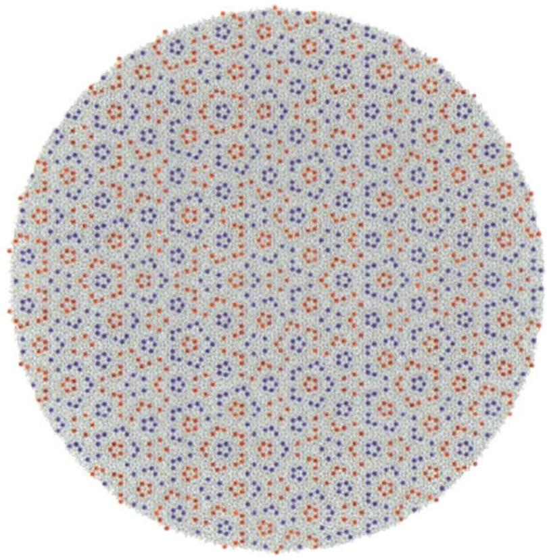

Figure 6.
The K-type vertex distribution plotted in two different colors, one for each of the distinct regions in the perpendicular space from where these configurations are projected.

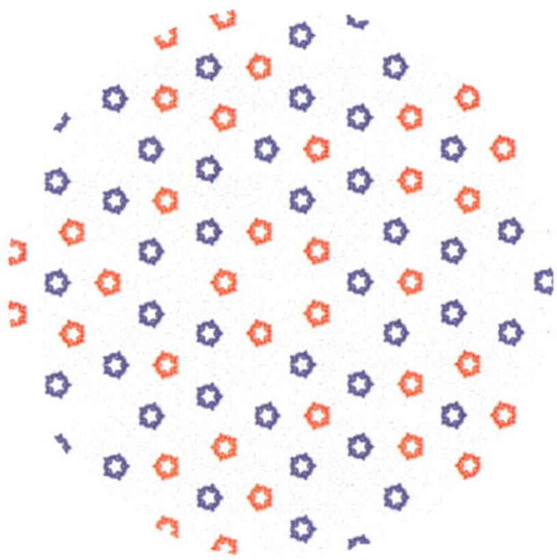

Figure 7.
Sun configurations surrounded by the five orientations of the K-type vertex plotted in two different colors corresponding to the two distinct regions in the perpendicular space from where these configurations are projected.

Figure 8.
Sun configurations surrounded by K vertex types from different regions in the perpendicular space displayed side-by-side. The black tiles represent the vertices with their empires turned on. The empires are colored in green.

shows a zoom-in region where only the sun configurations surrounded by K-type vertices are plotted also in different colors.

We have performed several studies of the empires of the K-type vertices that surround sun configurations. When analyzing just the K-sun configurations projected from the same region in the perpendicular space we are looking at the empire distributions, considering the empires from the K vertices. We consider one sun configuration to be completed, when all the K vertices surrounding it have their empires turned on. One of the interesting findings is that regardless of the number of completed suns—K vertex empires turned on—no other K vertex type surrounding a sun is covered by these empires. The K vertex patch tiles closest to the center—the sun—remain uncovered by the empires from the other suns. This is valid also for the case in which the sun configurations are projected from the other region in the perpendicular space. Provided these configurations are from the same region in the perpendicular space, there are some tiles in the K vertices that are not covered by empires coming from different suns—a "selective" nonlocality constrained by the higher dimensional representation. In **Figure 8**, we show the K-sun configurations side-by-side from both distinct regions in the perpendicular space.

5. Conclusions

In this chapter we have reviewed several properties of quasicrystals, their nonlocal empires, and the methods used to generate the quasicrystal configurations and the empires of their vertices. We have studied quasicrystals projected from higher dimensions, $\mathbb{Z}^5$ to 2D (the Penrose tiling), $\mathbb{Z}^6$ to $\mathbb{E}^3$ and $\mathbb{D}^6$ to $\mathbb{E}^3$ for the 3D case. For several vertex configurations, we have analyzed their empires—the nonlocal distribution of their forced tiles—in relation to the higher dimension representation. These nonlocal properties allow us to study the quasicrystal dynamics in a novel way, a nonlocal game-of-life approach, in which the empires and the possibility space dictate the movement and trajectory of the chosen quasiparticle configurations. Case studies of two-particle interactions based on nonlocal rules, while not exhaustive, are showing important similarities with other experimental physics discoveries, like time crystals. The research into the inherent nonlocality of quasicrystals proves very rich in describing the various quasicrystal configurations and their correlation with high dimensional representations. These studies open up a new, but very promising avenue of research that can bridge together different fields, like physics in high dimensions, gauge and group theory, phason dynamics and advanced material science.

Acknowledgements

We acknowledge the many discussions had with Richard Clawson about this project and we thank him for his useful comments and suggestions.

Author details

Fang Fang*†, Sinziana Paduroiu†, Dugan Hammock and Klee Irwin
Quantum Gravity Research, Los Angeles, CA, USA

*Address all correspondence to: fang@quantumgravityresearch.org

† These authors contributed equally.

References

[1] Shechtman D, Blech I, Gratias D, Cahn JW. Metallic phase with long-range orientational order and no translational symmetry. Physical Review Letters. 1984;**53**:1951. DOI: 10.1103/PhysRevLett.53.1951

[2] Kleinert H, Maki K. Lattice textures in cholesteric liquid crystals. Fortschritte der Physik. 1981;**29**:219259. DOI: 10.1002/prop.19810290503

[3] Levine D, Steinhardt PJ. Quasicrystals: A new class of ordered structures. Physical Review Letters. 1984;**53**:2477. DOI: 10.1103/PhysRevLett.53.2477

[4] Murr LE. Aperiodic crystal structures: Quasicrystals. Handbook of Materials Structures, Properties, Processing and Performance. Switzerland: Springer; 2014. pp. 1-9. DOI: 10.1007/978-3-319-01815-7

[5] Steurer W. Twenty years of structure research on quasicrystals. Part I. Pentagonal, octagonal, decagonal and dodecagonal quasicrystals. Zeitschrift für Kristallographie – Crystalline Materials. 2004;219:391-446. DOI: 10.1524/zkri.219.7.391.35643

[6] Hovmoller S, Hovmoller ZL, Zou X, Grushko B. Structures of pseudo-decagonal approximants in Al-Co-Ni. Philosophical Transactions of the Royal Society of London, Series A. 2012;**370**: 2949. DOI: 10.1098/rsta.2011.0310

[7] Singh D, Yun Y, Wan W, Grushko B, Zou X, Hovmoller S. A complex pseudo-decagonal quasicrystalapproximant, $Al_{37}(Co, Ni)_{15.5}$, solved by rotation electron diffraction. Journal of Applied Crystallography. 2014;**47**:215. DOI: 10.1107/S1600576713029294

[8] Singh D, Yun Y, Wan W, Grushko B, Zou X, Hovmoller S. Structure determination of a pseudo-decagonal quasicrystal approximant by the strong-reflections approach and rotation electron diffraction. Journal of Applied Crystallography. 2016;**49**:433-441. DOI: 10.1107/S1600576716000042

[9] Nagao K, Inuzuka T, Nishimoto K, Edagawa K. Experimental observation of quasicrystal growth. Physical Review Letters. 2015;**115**:075501. DOI: 10.1103/PhysRevLett.115.075501

[10] Grunbaum B, Shephard GC. Tilings and Patterns. New York: W. H. Freeman and Company; 1987. DOI: 0.1002/crat.2170260812

[11] Conway JH. Triangle tessellations of the plane. American Mathematical Monthly. 1965;**72**:915

[12] Penrose R. The role of aesthetics in pure and applied mathematical research. Bulletin of the Institute of Mathematics and Its Applications. 1974;**10**:266

[13] Effinger L. The empire problem in penrose tilings [thesis]. Williamstown US: Williams College; 2006

[14] Socolar JES, Steinhardt PJ. Quasicrystals. II. Unit-cell configurations. Physical Review B. 1986; **34**:617. DOI: 10.1103/PhysRevB.34.617

[15] Senechal MJ. Quasicrystals and Geometry. Cambridge: Cambridge University Press; 1995. DOI: 10.1002/adma.19970091217

[16] Fang F, Hammock D, Irwin K. Methods for calculating empires in quasicrystals. MDPI Crystals. 2017;**7**: 304. DOI: 10.3390/cryst7100304

[17] Fang F, Paduroiu S, Hammock D, Irwin K. Non-local game of life in 2D quasicrystals. MDPI Crystals. 2018;**8**: 416. DOI: 10.3390/cryst8110416

[18] Hammock D, Fang F, Irwin K. Quasicrystal tilings in three dimensions

and their empires. MDPI Crystals. 2018; **8**:370. DOI: 10.3390/cryst8100370

[19] Minnick L. Generalized forcing in aperiodic tilings [thesis]. Williams College; 1998

[20] Fang F, Irwin K. An icosahedral quasicrystal and E8 derived quasicrystals. arXiv:1511.07786. 2015

[21] Gardner M. Mathematical games the fantastic combinations of John Conways new solitaire game of 'life'. Scientific American. 1970;**223**:120

[22] de Bruijn NG. Algebraic theory of Penroses non-periodic tilings. Proceedings of the Koninklijke Nederlandse Akademie van Wetenschappen. 1981;**84**:53. DOI: 0.1016/1385-7258(81)90017-2

[23] Papadopolos Z, Kramer P, Zeidler D. The F-type icosahedral phase: Tiling and vertex models. Journal of Non-Crystalline Solids. 1993;**153**:215. DOI: 10.1016/0022-3093(93)90345-X

[24] Kramer P, Papadopolos Z, Zeidler D. Symmetries of icosahedral quasicrystals. In: Gruber B, Biedenharn LC, Doebner HD, editors. Symmetries in Science V. Boston: Springer; 2011. p. 395. DOI:10.1007/978-1-4615-3696-3_19

[25] Kramer P. Modelling of Quasicrystals. Physica Scripta. 1993; **T49**:343. DOI: 10.1088/0031-8949/1993/T49A/060

[26] Bailey DA, Lindsey KA. Game of Life on Penrose Tilings. arXiv:1708.09301. 2017

[27] Owens N, Stepney S. Investigations of Game of Life Cellular Automata Rules on Penrose Tilings: Lifetime and Ash Statistics. Automata-2008. Bristol: Luniver Press; 2008. p. 1

[28] Owens NDL, Stepney S. Investigations of game of life cellular automata rules on Penrose tilings: Lifetime, ash, and oscillator statistics. Journal of Cellular Automata. 2010;**5**: 207

[29] Wilczek F. Quantum time crystals. Physical Review Letters. 2012;**109**: 160401. DOI: 10.1103/PhysRevLett.109.160401

[30] Else DV, Nayak C. Classification of topological phases in periodically driven interacting systems. Physical Review B. 2016;**93**:201103. DOI: 10.1103/PhysRevB.93.201103

[31] Zhang J, Hess PW, Kyprianidis A, Becker P, Lee A, Smith J, et al. Observation of a discrete time crystal. Nature. 2017;**543**:217. DOI: 10.1038/nature21413

[32] Sacha K, Zakrzewski J. Time crystals: A review. Reports on Progress in Physics. 2018;**81**:016401. DOI: 10.1088/1361-6633/aa8b38

Chapter 4

Structure Analysis of Quasicrystal Approximants by Rotation Electron Diffraction (RED)

Devinder Singh and Sven Hovmöller

Abstract

Complete 3D electron diffraction can be collected by rotation electron diffraction (RED) for single-crystal powder-sized samples, i.e., <0.1 μm, in all dimensions. Data collection takes about 1 h and data processing takes another hour. The crystal structures are solved by standard crystallographic techniques. X-ray crystallography requires crystals several micrometers big. For nanometer-sized crystals, electron diffraction and electron microscopy (EM) are the only possibilities. Two methods have been developed for collecting complete (except for a missing cone) three-dimensional (3D) electron diffraction data: the rotation electron diffraction and automated electron diffraction tomography (ADT). By collecting 1000–2000 electron diffraction patterns, a complete 3D data set is obtained. The geometry in RED is analogous to the rotation method in X-ray crystallography; the sample is rotated continuously along one rotation axis. In recent years, large number of crystal structures has been solved by RED. These include the most complex zeolites ever solved and quasicrystal approximants, such as the pseudo-decagonal approximants PD2 and PD1 in Al-Co-Ni. In this chapter, the results of our recent studies on the structure analysis of complex pseudo-decagonal (PD) quasicrystal approximants PD2 (a = 23.2, b = 32.3, c = 4.1 Å) and PD1 (a = 37.3, b = 38.8, c = 4.1 Å) by RED have been discussed. These are known to be the most complicated approximant structures ever solved to atomic resolution by electron crystallography. PD2 and PD1 are built of characteristic 2 nm wheel clusters with fivefold rotational symmetry, which agrees with other approximants in the PD series as well as with the results from high-resolution electron microscopy images.

Keywords: electron crystallography, rotation electron diffraction, quasicrystal, approximant

1. Introduction

One of the most important techniques for studying crystals is electron crystallography. Recently, a new method, rotation electron diffraction (RED), has been developed for collecting three-dimensional (3D) electron diffraction data by combining electron beam tilt and goniometer tilt in a transmission electron microscope [1–4]. RED is capable of structure determination as well as phase identification of unknown crystals. It is easier, much faster, and more straightforward than powder X-ray diffraction and other electron microscopy techniques, such as high-resolution transmission electron microscopy (HRTEM). There is no enigma in the determination of unit cell, space group and indexing of diffraction peaks in RED [5].

The low-density structures such as zeolites and open-framework compounds are solved by RED method [5, 6]. Since, complex dense intermetallic compounds such as quasicrystal approximants contain heavy elements and thus suffer more from dynamical scattering, it is interesting to see if they can also be solved from RED data. Quasicrystals possess aperiodic long-range order associated with crystallographically forbidden rotational symmetries (5-, 8-, 10-, or 12-fold) and exhibit many outstanding physical properties [7–13]. Several breakthrough experiments performed by Dan Shechtman in 1982 on rapidly solidified Al-Mn alloys have led to the discovery of quasicrystals. It exhibits sharp diffraction peaks with icosahedral symmetry [14]. Quasicrystals exhibit unique physical properties which strongly differ from the properties of metals, insulators, and crystalline or amorphous phases [7–13, 15–21]. Thus, these materials have the potential to be used in many areas of advanced technology. Out of the many alloy systems which possess quasicrystalline phases, Al-based quasicrystalline alloy systems are easily available, cheap, and non-toxic.

The most critical aspect of quasicrystals from the experimental and theoretical point of view is to solve their structures. Their structures have been solved theoretically using a sequence of periodic structures with growing unit cells [22–24]. There exist also a number of crystalline phases resembling the quasicrystals, known as approximant phases [22]. The diffraction patterns of approximant phases are closely related to those of quasicrystals as their structures are built up by the same clusters as in quasicrystals. Quasicrystals and their approximant phases have similar electron diffraction patterns and chemical compositions [25–31], showing that they have similar local structures. One approach is to determine the structures of approximants. This helps us to get a deep understanding of the relationships between quasicrystals and their approximant phases. Thus, approximant phases may hold the key to determine the structures of quasicrystals.

The HRTEM and high-angle annular dark-field (HAADF) studies of Al-Co-Ni decagonal quasicrystals suggest that the basic structure is composed of 2 nm clusters with fivefold rotational symmetry [32–34]. A series of pseudo-decagonal (PD) quasicrystal approximants in Al-Co-Ni with almost 10-fold symmetry of their electron diffraction patterns have been found and described [35, 36]. Out of those approximants, only two structures, namely, PD4 [37] and PD8 (also called the W-phase) [38], have been solved to atomic resolution by X-ray crystallography. The PD1, PD2, PD3, and PD5 structures were solved at low resolution from the limited information provided by electron diffraction patterns, unit cell dimensions, and HRTEM images [39]. An attempt to solve the structures of PD1 and PD2 in $Al_{71}Co_{14.5}Ni_{14.5}$ alloy by maximum entropy Patterson deconvolution was reported by Estermann et al. [40]. Since these two structures were found to intergrow, thus there was a serious problem in the application of X-ray diffraction. This problem can be eliminated in the case of electron crystallography as much smaller crystals ($<1\ mm^3$) are needed for electron diffraction. Recently, we have solved the structures of PD2 and PD1 by RED method [41, 42]. The present chapter deals with the results and discussion of these two structures.

The decagonal quasicrystals are described by a quasiperiodic arrangement of clusters [43–46]. All decagonal quasicrystals in Al-Co-Ni and their high-order approximants are composed of 2 nm wheel clusters [47–51]. The arrangement of atoms within the clusters imposes restrictions on the cluster arrangements, e.g., an overlapping of clusters [52–56]. To understand the structure of quasicrystals, it is important to find the details of the atomic arrangements within the 2 nm wheel clusters and their packing into a 3D crystal. The geometrical building principles of

Al-Co-Ni, Al-Co-Cu, and Al-Fe-Ni decagonal quasicrystals and their approximant phases in terms of a fundamental unit cluster-based approach that leads to a unifying view of all these phases have been discussed [57]. This unit cluster has ~2 nm diameter.

The RED method has been applied for ab initio structure determination of PD2 (a = 23.2, b = 32.3, c = 4.1 Å) and PD1 (a = 37.3, b = 38.8, c = 4.1 Å), quasicrystal approximants in the Al–Co–Ni alloy system, and their structure determination by direct methods from the RED data set. After PD8 (a = 23.2, b = 19.8, c = 4.1 Å), PD2 has the second smallest unit cell area in the PD series [39]. a = 23.2 Å is the same as that for PD8, but b is τ (the golden mean 1.61803. ..) times larger than 19.8 Å, i.e., 32.0 Å, the same as that in PD4 (a = 101.3, b = 32.0, c = 4.1 Å). Compared with PD2 and PD8, PD1 has a larger unit cell and hence contains more atoms. Solving the structures of more complex quasicrystal approximants in the PD series from electron diffraction data by direct methods will be more challenging, because of the increased unit cell dimensions and number of unique atoms in the unit cell.

2. Materials and experimental procedure

The details of the preparation methods of $Al_{71}Co_{14.5}Ni_{14.5}$ nominal composition are reported elsewhere [41, 42]. Powder X-ray diffraction examination revealed a diffraction pattern typical of PDs [35]. A piece of the annealed sample was powdered and dispersed in ethanol and treated by ultrasonification for 2 min. A droplet of the suspension was transferred onto a copper grid (with carbon film). The 3D-RED data were collected on a JEOL JEM-2100 LaB6 microscope at 200 kV [1]. The single-tilt tomography sample holder was used for data collection. In RED, we combine electron beam tilt and goniometer tilt (**Figure 1**). The RED data collection software package was used which controls 3D-RED data collection in an automated way [1, 4, 58]. The selected area diffraction patterns were collected at each tilt angle from a μm-sized crystal (**Figure 1(b)**). For RED data collection, electron beam tilt with many small steps and goniometer tilt with larger steps was combined to cover

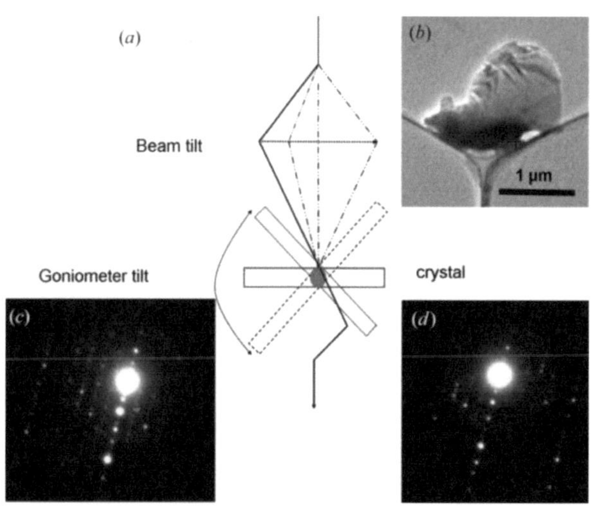

Figure 1.
(a) Ray diagram of the electron beam rotation, showing beam tilt and goniometer tilt. (b) a single crystal of size ~2.0 × 1.0 × <0.1 mm was used for the RED data collection. (c) and (d) two diffraction patterns of PD2 which are 1.0° (20 frames) apart. Reproduced with permission of the International Union of Crystallography (https://scripts.iucr.org/cgi-bin/paper?HE5621) [41].

a large part of reciprocal space. **Table 1** gives the details of the RED data collection and crystallographic information for the PD2 and PD1 quasicrystal approximants. Energy-dispersive spectroscopy (EDS) analysis was carried out on the same crystal after the RED data collection which showed that the composition was close to the nominal one.

The software package RED data processing was used for the data processing of the collected frames [4, 58], including direct beam-shift correction, peak search, unit cell determination, indexing of reflections, and intensity extraction. ED frames collected were combined into a 3D data set for reciprocal space reconstruction. After reciprocal space had been reconstructed, the unit cell parameters, space group, reflection indices, and diffraction intensities were determined. The indexing

Name	**Pseudo-decagonal (PD2) quasicrystal approximant**	**Pseudo-decagonal (PD1) quasicrystal approximant**
Chemical formula	$Al_{37}(Co,Ni)_{15.5}$	$Al_{77}(Co/Ni)_{31}$
Temperature (K)	298	298
Wavelength (Å)	0.02508	0.02508
Crystal system	Orthorhombic	Orthorhombic
Space group	*Pnmm*	*Pnam*
Unit cell parameters (Å)	a = 23.2, b = 32.3, c = 4.1	a = 37.3, b = 38.8, c = 4.1
Volume (Å^3)	3075.7	5933.68
Density (calculated in Mg cm^{-3})	4.132	4.374
Crystal size (μm)	2.0 × 1.0 × <0.1	2.0 × 2.0 × <0.1
Tilt range (°)	−74.3 to +36.0	+29.5 to −64.6
Tilt step (°)	0.05	0.05
Exposure time/frame (s)	0.5	0.2
Total data collection time (min)	90	90
No. of frames	2255	2050
Program for structure determination	*SHELX97*	*SHELX97*
Resolution (Å)	1.0	1.0
Completeness (%)	89.3	94.5
Reflections collected	8153	7070
R(int)	0.33	0.26
Observed unique reflections ($I > 2\sigma$)	1799	2588
Parameters/restraints	156 with 0 restraint	325 with 0 restraint
Goodness-of-fit on F^2	4.155	2.854
Final R indices ($I > 2\sigma$)	R_1 = 0.4285, wR_2 = 0.7023	R_1 = 0.3606, wR_2 = 0.6641
R (all reflections)	0.4326	0.3671
Highest peak and deepest hole	1.98 and – 2.56	1.31 and – 1.42

Reproduced with permission of the International Union of Crystallography (https://scripts.iucr.org/cgi-bin/paper?HE5621 & https://scripts.iucr.org/cgi-bin/paper?jo5016) [41, 42].

Table 1.
Crystallographic data, RED experimental parameters, and structure refinement details for the PD2 and PD1 quasicrystal approximant structures.

of all reflections has been done. For the determination of space group, the two-dimensional slices cut from the 3D-RED data along the ($hk0$), ($h0l$), and ($0kl$) planes were used to derive the extinction conditions. The final file produced with *hkl* intensity was used for solving the structure by standard crystallographic techniques. Based on RED intensities, the structure was solved by direct methods and refined using the program *SHELX*97 [59, 60]. Nearly all atoms could be located and refined isotropically using the RED data. The simulated electron diffraction patterns were calculated using the intensities obtained from the output of *SHELX*97.

3. Structure analysis of PD2 and PD1 quasicrystal approximants

3.1 RED data processing

The 3D reciprocal space can be obtained by combining the series of electron diffraction frames. RED data processing program is used for the reciprocal space reconstruction of the electron diffraction data. The unit cell dimensions for PD2 and PD1 were found to be a = 23.2, b = 32.3, c = 4.1 Å, angles α = 89.7, β = 90.1, γ = 89.3° (**Figure 2**) and a = 37.3, b = 38.8, c = 8.2 Å, α = 89.9, β = 90.1, γ = 90.1°, respectively, indicating that both PD2 and PD1 are orthorhombic (**Table 1**). We present here only the RED images of the PD2 quasicrystal approximant. For the RED images of PD1, we refer the readers to the reference [42]. The entire 3D reciprocal lattice of PD2 obtained from the 3D-RED data viewed along c^* is shown in **Figure 2(a)**. Only the data out to 1.0 Å are shown because the reflections outside 1.0 Å were too weak to be detected and the completeness was too low. The presence of several 10-fold rings can be seen. The c lattice parameter is described as either 4.1 or 8.2 Å. The longer c-axis dimension (8.2 Å) can be considered as the cell parameter of a superstructure.

Figure 2(b) shows the original data set projected along b^*; the 8.2 Å layers are seen. The odd layers (corresponding to the 8.2 Å c axis, shown in red) are much weaker than the even layers (corresponding to a 4.1 Å c axis). For a = 46.4, b = 64.6, and c = 8.2 Å, the space group was found to be *F*222 (No. 22), *F*2*mm* (No. 42), or *Fmmm* (No. 69). However, it is possible to treat the structure as having a c-axis of 4.1 Å. Since the total intensities of reflections with even l indices are more than four

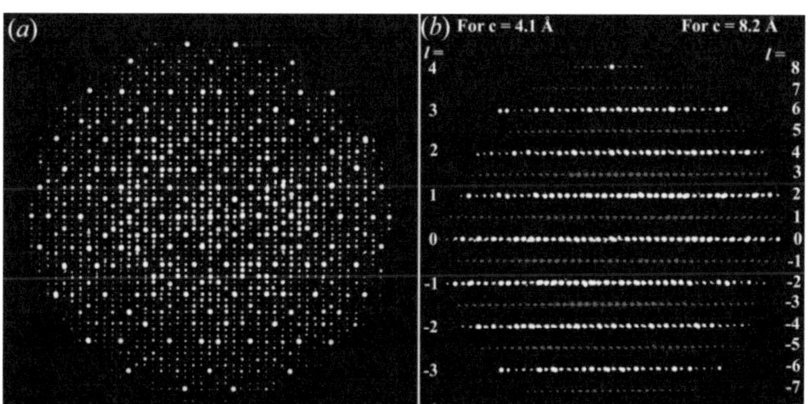

Figure 2.
(a) The entire 3D reciprocal lattice of PD2 obtained from the 3D-RED data viewed along c. (b) the original data set projected along b*. the odd layers (corresponding to the 8.2 Å c-axis, shown in red) are much weaker than the even layers (corresponding to a 4.1 Å c axis). Reproduced with permission of the International Union of Crystallography (https://scripts.iucr.org/cgi-bin/paper?HE5621) [41].*

times higher than those with odd l indices for c = 8.2 Å; the basic structure, i.e., using c = 4.1 Å, has been solved by only considering the reflections of even l indices. The axes a and b are selected in such a way so that the diffraction spot present at 2.0 Å resolution is along b^* and the equally strong diffraction spot present at 2.3 Å resolution is along a^* (**Figure 3**).

Figure 4(a)–(c) show the 2D slices ($hk0$), ($h0l$), and ($0kl$) of the reconstructed reciprocal lattice obtained from the 3D-RED data. Each slice corresponds to one complete quadrant for orthorhombic compounds containing all unique reflections. The missing reflections attributed to the missing cone. The odd layers (corresponding to the 8.2 Å c-axis, shown in red colour in (b) and (c)) are much weaker than the even layers (corresponding to a 4.1 Å c-axis).

Figure 5(a)–(c) show 2D slices of ($hk1$) and ($hk-1$), ($hk2$) and ($hk-2$), and ($hk3$) and ($hk-3$) corresponding to c = 4.1 Å. Two layers of each are combined and shown together. The white reflections correspond to hkl, while yellow corresponds to $hk-l$ layers. The corresponding calculated kinematical electron diffraction patterns agree very well (**Figure 5(d)–(f)**). Notice the presence of many rings of 10 strong reflections. This is typical of 10-fold quasicrystals and their approximants.

3.2 Solving the basic atomic structure and deducing an atomic model of PD2 and PD1

Based on the systematic absences for the unit cell with c = 4.1 Å for PD2 ($0kl$: $k + l = 2n + 1$; $0\,k0$: $k = 2n + 1$; $00\,l$: $l = 2n + 1$) and PD1 ($0kl$: $k + l = 2n$; $h0l$: $h = 2n$; $h00$: $h = 2n$; $0\,k0$: $k = 2n$; $00\,l$: $l = 2n$), the space groups were found to be [$Pnmm$ (No. 59) or $Pnm2_1$ (No. 31)] for PD2 and [$Pnam$ (No. 62) or $Pna2_1$ (No.33)] for PD1. $Pnmm$ and $Pnam$ are centrosymmetric and have a higher symmetry than $Pnm2_1$ and $Pna2_1$. Since most inorganic structures are centrosymmetric, thus the PD2 and PD1 structures were first solved in $Pnmm$ and $Pnam$, respectively.

Since the procedure followed for the structure solution and refinement using RED data is same for both the structures, we discuss here only the step-by-step

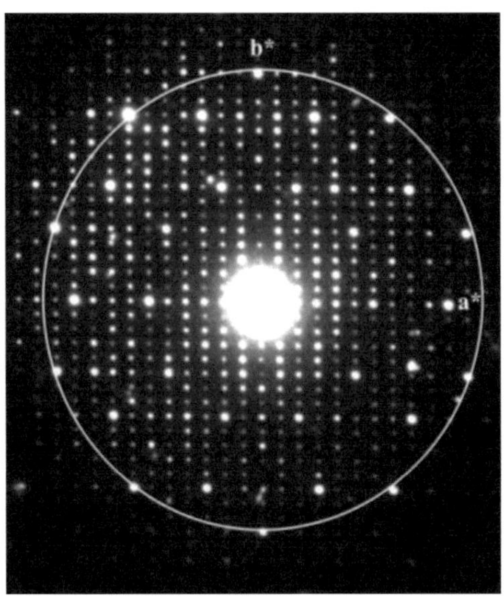

Figure 3.
Selected area electron diffraction pattern of the hk0 layer of PD2. This electron diffraction pattern was collected on a JEOL JEM-2000FX microscope. Reproduced with permission of the International Union of Crystallography (https://scripts.iucr.org/cgi-bin/paper?HE5621) [41].

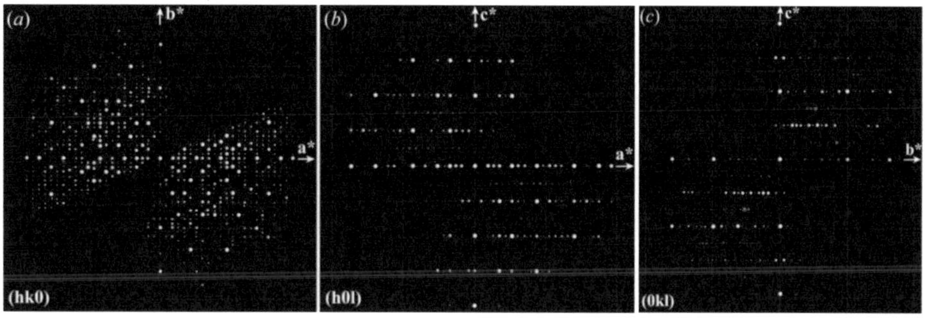

Figure 4.
Two-dimensional slices of the reconstructed reciprocal lattice obtained from the 3D-RED data. (a) (hk0), (b) (h0l), and (c) (0kl). The layers shown in red colour in (b) and (c) for c = 8.2 Å are much weaker than the even layers of c = 4.1 Å. reproduced with permission of the International Union of Crystallography (https://scripts.iucr.org/cgi-bin/paper?HE5621) [41].

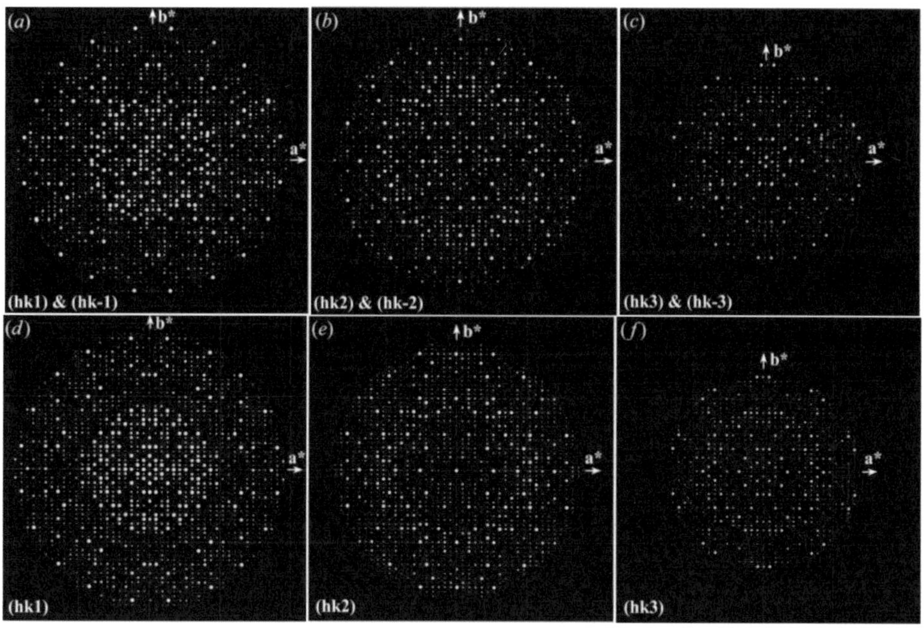

Figure 5.
(a)–(c) 2D slices of (hk1) and (hk − 1), (hk2) and (hk − 2), and (hk3) and (hk-3) for c = 4.1 Å obtained from experimental RED data. Here, two layers of each are combined and shown together. The white reflections correspond to hkl while yellow corresponds to hk − l layers. (d)–(f) simulated kinematical electron diffraction patterns after the final refinement of the structure model using c = 4.1 Å [(hk1), (hk2) and (hk3), respectively]. Reproduced with permission of the International Union of Crystallography (https://scripts.iucr.org/cgi-bin/paper?HE5621) [41].

details in the structure determination of PD2 structure. The details for the PD1 structure is reported elsewhere [42]. The crystallographic data, RED experimental parameters, and structure refinement details for the PD2 and PD1 structures are given in **Table 1**. In the case of PD2, a total of 8153 reflections, of which 1799 are unique, within 1.0 Å resolution, were collected. The structure model of PD2 was deduced by direct methods using *SHELX97*. The refinement of structure was done by taking the square root of the intensities as an estimate for the standard deviation (σ). The final structure refinement with the 3D-RED data converged to $R_1 = 0.43$ for the 1799 unique reflections (89.3% of all unique reflections up to $d \geq 1.0$ Å were observed above the background noise level). From the structure solution, the first 26 highest unique peaks (atoms) found by *SHELX* were examined. Two 2 nm wheels were clearly seen per unit cell. There are three concentric rings of atoms in each wheel: innermost 5, then 10, and finally 20 arranged in 10 pairs. These three rings comprise of total 17 unique atoms. These 17 atoms are considered as Co/Ni atoms. In the periodic table, Co and Ni lie adjacent to each other and thus cannot be distinguished by the present technique. The nine remaining peaks were assigned to Al. They can be seen at several places in the unit cell but did not form patterns of fivefold symmetry. In one 2 nm wheel cluster, the five innermost atoms have the same z coordinate (in *Pnmm,* because of the short c-axis, all atoms must be located at one of the two mirror planes $z = 0.25$ or $z = 0.75$). In the next ring the 10 Co/Ni atoms are arranged in an alternate manner at $z = 0.25$ and 0.75 (**Figure 6(c)**). At the rim of the 2 nm wheel cluster, the 20 atoms are arranged in pairs when viewed along the c-axis. One atom is at $z = 0.25$ and the other at $z = 0.75$ in each such pair. For any two such pairs, the nearest atoms in adjacent pairs are at the same height, i.e., either both are at $z = 0.25$ or both are at $z = 0.75$. The assigned 17 atoms as Co/Ni correspond to the 14 highest peaks and peaks 16, 19, and 21. Thus only four of the highest peaks, i.e., 15, 17, 18 and 20, were considered to arise from Al.

Comparing the structure model of PD2 generated by *SHELXS97* with that of PD4, which was obtained from single-crystal X-ray diffraction [37], we found a one-to-one agreement for all the 35 Co/Ni atoms in the 2 nm wheels. This similarity is not limited to the projected structure; all the coordinates of z also agree between PD2 and PD4. In the region of small hour glass shaped between two

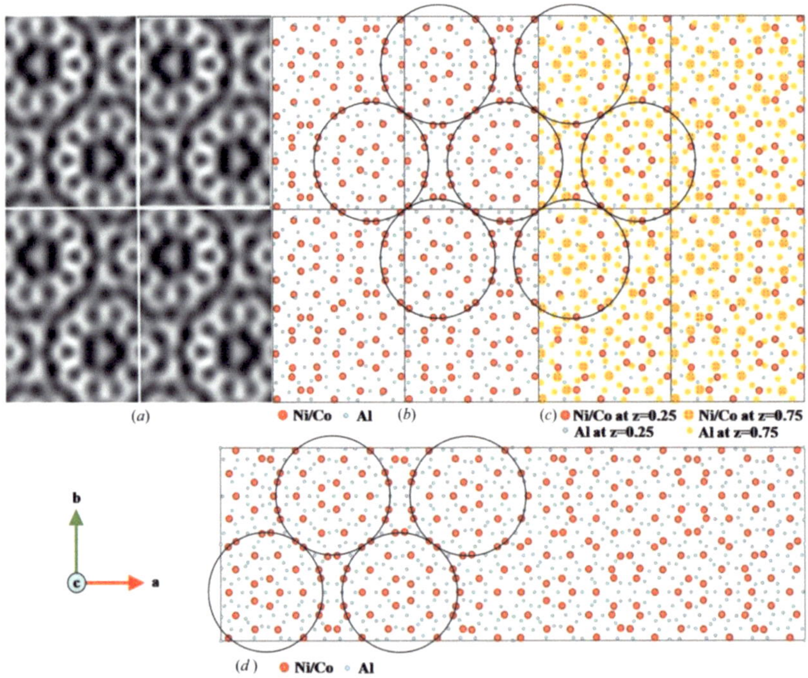

Figure 6.
(a) HRTEM image of PD2 with 2 × 2 unit cells taken along the c axis [39]. The 10-fold wheels containing atoms (in black color) are clearly seen. (b) the PD2 atomic structure model projected along the c-axis. The 2 nm wheel cluster columns with pseudo 10-fold rotational symmetry are shown by circles. Co/Ni atoms are shown in red color while Al atoms in blue. (c) the Co/Ni atoms at z = 0.25 (red) and z = 0.75 (red with yellow cross) layers are shown in the structure model. (d) the structure of PD4 (a = 101.3, b = 32.0, c = 4.1 Å) as determined by X-ray crystallography [37]. The circles mark the 2 nm clusters similar to those found in PD2. Reproduced with permission of the International Union of Crystallography (https://scripts.iucr.org/cgi-bin/paper?HE5621) [41].

non-intersecting wheels, there were no Co/Ni atoms. After finding all the 35 Co/Ni atoms, we started observing for the Al atoms among the Q peaks in the difference Fourier maps generated by *SHELX*. After each refinement step, the stability of the atoms in the structure model was checked. The atomic displacements after the refinement did not show significant movements. We have found 55 unique atomic positions (17 Co/Ni and 38 Al) in the unit cell with a reasonable geometry after refinement using *SHELXL97*. This is close to the nominal and experimentally determined (by EDS) chemical composition except that one or two Al atoms may still be missing from our model. The structure model was also deduced using the strong reflections approach [61] which is quite similar to that obtained by direct methods. **Figure 6(b)** and **(c)** shows the structure model of PD2 obtained from the RED data. The high value of R_1 (43%) is normal for data obtained from electron diffraction. This high value may come from twins and intergrowth with other approximants in the PD series, especially for PD1. It partly also arises due to the distortions of the intensities by multiple scattering. The PD4 also had a remarkably high R-value (24.5%), even though it was solved from single-crystal X-ray diffraction data. After the final refinement of PD2, the chemical composition calculated was found to be $Al_{37}(Co/Ni)_{15.5}$. The atomic structure shown here agrees well with the lower resolution structure model obtained from HRTEM images [39]. **Figure 6(a)** shows an experimental HRTEM image of PD2 taken along the *c*-axis after applying crystallographic image processing using *CRISP* [62]. The plane group symmetry was found to be *pgg* for PD2. The transition metal atoms are black in color with 10-fold wheels are clearly seen. The wheels are 23.2 Å apart (along *a*), while vertically (along *b*) they are 32.2 Å apart, and they are intersecting diagonally at 19.8 Å. As shown in **Figure 6**, the projected unit cell consists of circular wheel clusters of 2 nm in diameter. Around the perimeters of each of the 2 nm cluster columns, dark spots belonging to Co and Ni atoms appear.

Figure 6(c) shows the atomic arrangement in the z = 0.25 and z = 0.75 layers. Some of the atoms which appear to be very close to each other in projection are actually separated by 2.05 Å along z. The 5 Co/Ni and 5 Al atoms which form the pentagons in the circular wheel cluster are present in different layers. Outside the pentagon, there are three circular arrangements of atoms. There are 10 Al and 10 Co/Ni atoms in the first and second circles, respectively. The atoms present in these circles at z = 0.25 are followed by atoms at z = 0.75 and vice versa. Thus, this leads to the formation of systematic circular sequence of atoms present in different layers. The third circle consists of 10 pairs of Co/Ni atoms. The intersecting wheels share four Co/Ni atoms from two pairs, thus locking in the wheel positions relative to each other. Simulated electron diffraction patterns are produced using the intensities found from the output of *SHELX97* after the final refinement for the structure model. The simulated electron diffraction patterns generated are found to be in good agreement with experimental electron diffraction patterns (**Figure 5**). The simulated electron diffraction patterns show the presence of 10-fold symmetry.

A total of 7070 reflections were collected for the PD1 structure. Out of which 2588 are unique. The data completeness is 94.5% for the reflections with d ≥ 1.0 Å. The R_{int} value is found to be 0.26, which is much higher than that for single-crystal X-ray diffraction but normal for electron diffraction data. The causes for this relatively poor data quality compared to single-crystal X-ray diffraction are currently under investigation. The program *SHELXL* was used for structure refinement. The final structure refinement for the 3D-RED data converged to R_1 = 0.36 without using any geometric restraints. **Figure 7(b)** shows the structure model of PD1 obtained from RED data. **Table 1** gives the crystallographic data, RED experimental parameters, and structure refinement details for the PD1 structure model. Successive refinement using *SHELXL* gives in 108 unique atomic positions

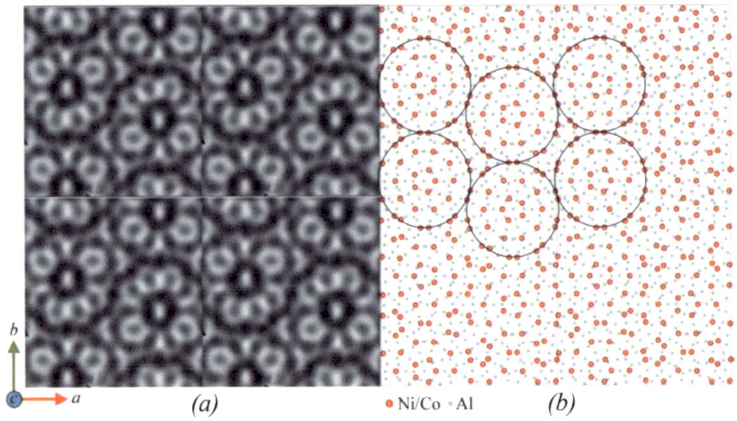

Figure 7.
(a) HRTEM image of PD1 with 2 × 2 unit cells, taken along the c axis [39]. Transition metal atoms appear as black regions around the perimeters of each of the 2 nm wheel clusters. (b) Atomic structure model of PD1 obtained by RED after final refinement, projected along the c-axis. Reproduced with permission of the International Union of Crystallography (https://scripts.iucr.org/cgi-bin/paper?jo5016) [42].

(31 Co/Ni and 77 Al) all with a reasonable geometry. After the refinement of PD1, the chemical composition calculated was found to be $Al_{77}(Co/Ni)_{31}$, which is close to the nominal and experimental chemical composition determined by EDS, except that a few Al atoms may still be missing from our model. The PD1 structure (108) has almost twice as many unique atoms as that of PD2 (55). The structural model presented here agrees well with the experimental HRTEM image of PD1 [39]. The HRTEM image of PD1 (taken along the *c*-axis) after applying crystallographic image processing using *CRISP* is shown in **Figure 7(a)**. The plane-group symmetry was found to be *pgg* for PD1. The 2 nm wheels are clearly seen. As shown in **Figure 8(a)** and **(b)** for PD1 and PD2, the atoms present at the layers $z = 0.25$ and $z = 0.75$ form systematic circular sequences. At the center of each wheel, five Co/Ni

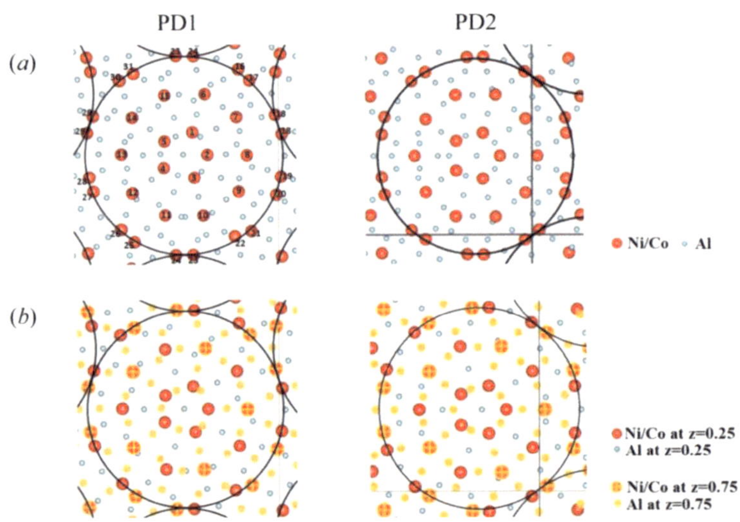

Figure 8.
(a) Circular wheel clusters of PD1 with PD2 obtained from RED data are compared. (b) PD1 and PD2 show identical arrangements of Ni/Co atoms present at z = 0.25 (red) and z = 0.75 (red with yellow cross) within the wheel cluster. Although, most of the Al atoms appear in similar locations within the wheels in PD1 and PD2, there are some differences. Reproduced with permission of the International Union of Crystallography (https://scripts.iucr.org/cgi-bin/paper?jo5016) [42].

and five Al atoms present in different layers form pentagons. There are four circular arrangements of atoms outside these pentagons. There are 10 Al atoms in the first circle, 10 Co/Ni and nearly 20 Al atoms in the second circle, nearly 30 Al atoms in the third circle, and 10 pairs of Co/Ni atoms and just a few Al atoms in the fourth and outermost ring. In all these circles, the atoms present at z = 0.25 are followed by atoms at z = 0.75 in a very regular fashion quite similar in PD1 and PD2.

The positions for all stronger Co/Ni scatterers are correct, while the positions of weaker Al scatterers are more uncertain. As discussed earlier, few Al atoms may be missing, few may be misplaced, and several may have split occupancies or could be shared Al/Co and/or Al/Ni sites. With the present data quality of electron diffraction, such fine details cannot be determined unambiguously. Some work has been done and some in progress on the ways to compensate the problems with respect to quality of data and absorption that combine to give electron diffraction intensity data that are inferior to those collected by X-ray diffraction. In the present case, the structure refinement can be done, and at least the Co/Ni atoms were found to be stable during refinement. The arrangement of Co/Ni atoms is in excellent agreement with previous studies by single-crystal X-ray diffraction for PD8 [38] and PD4 [37] and with the low-resolution projections obtained by HRTEM on PD1 [39].

4. Conclusions

Based on the results described and discussed in this chapter, it is proven that rotation electron diffraction method is an effective method to solve the structures of a rather complex and dense quasicrystal approximants. The structural details of pseudo-decagonal (PD) quasicrystal approximants PD2 and PD1 discussed in this chapter helped us to understand the atomic arrangements within the 2 nm wheel clusters. These are one of the most complex structures ever solved to atomic resolution by electron diffraction. The structural models obtained from the RED data agree well with the high-resolution transmission electron microscopy images.

Acknowledgements

One of the author (D. Singh) gratefully acknowledges the financial support by Department of Science and Technology (DST), New Delhi, India, in the form of INSPIRE Faculty Award [IFA12-PH-39]. The authors thank the Swedish Research Council (VR), the Swedish Governmental Agency for Innovation Systems (VINNOVA), and the Knut and Alice Wallenberg Foundation for the financial support through the project grant 3DEM-NATUR.

Author details

Devinder Singh[1*] and Sven Hovmöller[2]

1 Amity School of Applied Sciences, Amity University, Lucknow, UP, India

2 Department of Materials and Environmental Chemistry, Stockholm University, Stockholm, Sweden

*Address all correspondence to: dsingh2@lko.amity.edu

References

[1] Zhang D, Oleynikov P, Hovmöller S, Zou XD. Collecting 3D electron diffraction data by the rotation method. Zeitschrift fuer Kristallographie. 2010;**225**:94-102

[2] Zou XD, Hovmöller S, Oleynikov P. Electron Crystallography: Electron Microscopy and Electron Diffraction. Oxford, England, UK: International Union of Crystallography, Oxford Science Publications; 2011. ISBN: 978-0-19-958020-0, Chapter 11

[3] Willhammar T, Sun J, Wan W, Oleynikov P, Zhang D, Zou XD, et al. Structure and catalytic properties of the most complex intergrown zeolite ITQ-39 determined by electron crystallography. Nature Chemistry. 2012;**4**:188-194

[4] Wan W, Sun J, Su J, Hovmöller S, Zou XD. Three-dimensional rotation electron diffraction: Software RED for automated data collection and data processing. Journal of Applied Crystallography. 2013;**46**:1863-1873

[5] Martínez-Franco R, Moliner M, Yun Y, Sun J, Wan W, Zou XD, et al. Synthesis of an extra-large molecular sieve using proton sponges as organic structure-directing agents. Proceedings of the National Academy of Sciences of the United States of America. 2013;**110**:3749-3754

[6] Su J, Kapaca E, Liu L, Georgieva V, Wan W, Sun J, et al. Structure analysis of zeolites by rotation electron diffraction (RED). Microporous and Mesoporous Materials. 2013;**189**:115-125

[7] Yadav TP, Singh D, Tiwari RS, Srivastava ON. Enhanced microhardness of mechanically activated carbon–quasicrystal composite. Materials Letters. 2012;**80**:5-8

[8] Singh D, Mandal RK, Tiwari RS, Srivastava ON. Nanoindentation characteristics of $Zr_{69.5}Al_{7.5-x}Ga_xCu_{12}Ni_{11}$ glasses and their nanocomposites. Journal of Alloys and Compounds. 2011;**509**:8657-8663

[9] Singh D, Yadav TP, Mandal RK, Tiwari RS, Srivastava ON. Effect of Ga substitution on the crystallization behaviour and glass forming ability of Zr–Al–Cu–Ni alloys. Materials Science and Engineering A. 2010;**527**:469-473

[10] Singh D, Yadav TP, Mandal RK, Tiwari RS, Srivastava ON. Indentation characteristics of metallic glass and nanoquasicrystal-glass composite in Zr–Al(Ga)–Cu–Ni alloys. Intermetallics. 2010;**18**:2445-2452

[11] Singh D, Yadav TP, Mandal RK, Tiwari RS, Srivastava ON. Effect of Ti addition on the quasicrystalline phase formation and indentation characteristics of $Zr_{69.5}Al_{7.5}Cu_{12}Ni_{11}$ alloy. Philosophical Magazine. 2011;**91**:2837

[12] Singh D, Singh D, Yadav TP, Mandal RK, Tiwari RS, Srivastava ON. Synthesis and indentation behavior of amorphous and nanocrystalline phases in rapidly quenched Cu–Ga–Mg–Ti and Cu–Al–Mg–Ti alloys. Metallography, Microstructure, and Analysis. 2013;2:321-327

[13] Singh D, Yadav TP, Mandal RK, TiwariRS,SrivastavaON.Nanoindentation studies of metallic glasses and nanoquasicrystal glass composites in ZrAl(Ga)CuNi alloys. International Journal of Nanoscience. 2011;**10**:929-933

[14] Shechtman D, Blech I, Gratias D, Cahn J. Metallic phase with long-range orientational order and no translational symmetry. Physical Review Letters. 1984;**53**:1951-1953

[15] Somekawa H, Watanabe H, Mukai T. Damping properties in

Mg–Zn–Y alloy with dispersion of quasicrystal phase particle. Materials Letters. 2011;**65**:3251-3253

[16] Dubois J. So useful, those quasicrystals. Israel Journal of Chemistry. 2011;**51**:1168-1175

[17] Singh D, Shahi RR, Yadav TP, Mandal RK, Tiwari RS, Srivastava ON. Hydrogenation of $(Zr_{69.5}Al_{7.5}Cu_{12}Ni_{11})_{100-x}Ti_x$ quasicrystalline alloys and its effect on their structural and microhardness behaviour. Journal of Non-Crystalline Solids. 2013;**380**:11

[18] Yadav TP, Singh D, Shahi RR, Shaz MA, Tiwari RS, Srivastava ON. Formation of quasicrystalline phase in $Al_{70-x}Ga_xPd_{17}Mn_{13}$ alloys. Philosophical Magazine. 2011;**91**:2474

[19] Singh D, Singh D, Mandal RK, Srivastava ON, Tiwari RS. Crystallization behavior and mechanical properties of $(Al_{90}Fe_5Ce_5)_{100-x}Ti_x$ amorphous alloys. Journal of Alloys and Compounds. 2016;**687**:990-998

[20] Yadav TP, Singh D, Shaz MA, Tiwari RS, Srivastava ON. Synthesis of quasicrystalline film of Al–Ga–Pd–Mn alloy. Thin Solid Films. 2013;**534**:265

[21] Singh D, Singh D, Srivastava ON, Tiwari RS. Microstructural effect on the low temperature transport properties of Ce–Al (Ga) metallic glasses. Scripta Materialia. 2016;**118**:24-28

[22] Goldman A, Kelton R. Quasicrystals and crystalline approximants. Reviews of Modern Physics. 1993;**65**:213-230

[23] Cahn JW, Gratias D, Shechtman D. Pauling's model not universally accepted. Nature. 1986;**319**:102-103

[24] Duneau M, Mosseri R, Oguey C. Approximants of quasiperiodic structures generated by the inflation mapping. Journal of Physics A. 1989;**22**:4549-4564

[25] Singh D, Tiwari RS, Srivastava ON. Phase formation in rapidly quenched Cu-based alloys. Journal of Materials Science. 2009;**44**:3883-3888

[26] Bancel PA. Dynamical phasons in a perfect quasicrystal. Physical Review Letters. 1989;**63**:2741-2744

[27] Abe E, Tsai A. Quasicrystal-crystal transformation in Zn–Mg-rare-earth alloys. Physical Review Letters. 1999;**83**:753-756

[28] Zou X, Fung K, Kuo K. Orientation relationship of decagonal quasicrystal and tenfold twins in rapidly cooled Al–Fe alloy. Physical Review B. 1987;**35**:4526-4528

[29] Singh D, Hovmöller S, Grushko B, Wan W, Yun Y, Zou XD. A complex pseudo-decagonal quasicrystal approximant solved by strong reflections approach. Acta Crystallographica Section A. 2017;**A73**:c1192

[30] Hovmöller S, Singh D, Grushko B, Wan W, Yun Y, Zou XD. A complex pseudo-decagonal quasicrystal approximant solved by the strong reflections approach and refined against rotation electron diffraction (RED) data. Acta Crystallographica Section A. 2016;**A72**:s318

[31] Hovmöller S, Singh D, Wan W, Yun Y, Wan W, Grushko B, et al. Quasicrystal approximants solved by rotation electron diffraction (RED). Acta Crystallographica Section A. 2014;**A70**:c1195

[32] Hiraga K, Sun W, Ohsuna T. Structure of a pentagonal quasicrystal in $Al_{72.5}Co1_{7.5}Ni_{10}$ studied by high-angle annular detector dark-field scanning transmission electron microscopy. Materials Transactions, JIM. 2001;**42**:1146-1148

[33] Hiraga K, Ohsuna T, Yubuta K, Nishimura S. The structure of an

Al–Co–Ni crystalline approximant with an ordered arrangement of atomic clusters with pentagonal symmetry. Materials Transactions, JIM. 2001;**42**:897-900

[34] Saitoh K, Tsuda K, Tanaka M. New structural model of an $Al_{72}Ni_{20}Co_8$ decagonal quasicrystal. Journal of the Physical Society of Japan. 1998;**67**:2578-2581

[35] Grushko B, Holland-Moritz D, Wittmann R, Wilde G. Transition between periodic and quasiperiodic structures in Al–Ni–Co. Journal of Alloys and Compounds. 1998;**280**:215-230

[36] Döblinger M. PhD thesis. Berlin, Germany: University of Karlsruhe, Logos Verlag; 2002

[37] Oleynikov P, Demchenko L, Christensen J, Hovmöller S, Yokosawa T, Döblinger M, et al. Structure of the pseudodecagonal Al–Co–Ni approximant PD4. Philosophical Magazine. 2006;**86**:457-462

[38] Sugiyama K, Nishimura S, Hiraga K. Structure of a W–(AlCoNi) crystalline phase related to Al–Co–Ni decagonal quasicrystals, studied by single crystal X-ray diffraction. Journal of Alloys and Compounds. 2002;**342**:65-71

[39] Hovmöller S, Zou L, Zou XD, Grushko B. Structures of pseudo-decagonal approximants in Al–Co–Ni. Philosophical Transactions of the Royal Society of London. Series A. 2012;**370**:2949-2959

[40] Estermann MA, Lemster K, Steurer W, Grushko B, Döblinger M. Structure solution of a high-order decagonal approximant $Al_{71}Co_{14.5}Ni_{14.5}$ by maximum entropy Patterson deconvolution. Ferroelectrics. 2001;**250**:245-248

[41] Singh D, Yun Y, Wan W, Grushko B, Hovmöller S, Zou XD. A complex pseudo-decagonal quasicrystal approximant $Al_{37}(Co,Ni)_{15.5}$ solved by the rotation electron diffraction (RED) method. Journal of Applied Crystallography. 2014;**47**:215

[42] Singh D, Yun Y, Wan W, Grushko B, Hovmöller S, Zou XD. Structure determination of a pseudo-decagonal quasicrystal approximant by the strong-reflections approach and rotation electron diffraction. Journal of Applied Crystallography. 2016;**49**:433-441

[43] Guyot P, Audier M. A quasicrystal structure model for Al–Mn. Philosophical Magazine B. 1985;**52**:L15-L19

[44] Elser V, Henley CL. Crystal and quasicrystal structures in Al–Mn–Si alloys. Physical Review Letters. 1985;**55**:2883-2886

[45] Steurer W. Five-dimensional Patterson analysis of the decagonal phase of the system Al–Mn. Acta Crystallographica Section B. 1989;**45**:534-542

[46] Jeong HC, Steinhardt PJ. Cluster approach for quasicrystals. Physical Review Letters. 1994;**73**:1943-1946

[47] Hiraga K, Lincoln FJ, Sun W. Structure and structural change of Al–Ni–Co decagonal quasicrystal by high-resolution electron microscopy. Materials Transactions, JIM. 1991;**32**:308-314

[48] Saitoh K, Tsuda K, Tanaka M, Tsai AP, Inoue A, Masumoto T. Electron microscope study of the symmetry of the basic atom cluster and a structural change of decagonal quasicrystals of Al–Cu–Co alloys. Philosophical Magazine A. 1996;**73**:387-398

[49] Saitoh K, Tsuda K, Tanaka M, Kaneko K, Tsai AP. Structural study of an $Al_{72}Ni_{20}Co_8$ decagonal quasicrystal

using the high-angle annular dark-field method. Japanese Journal of Applied Physics. 1997;**36**:L1400-L1402

[50] Tsuda K, Nishida Y, Saitoh K, Tanaka M, Tsai AP, Inoue A, et al. Structure of Al–Ni–Co decagonal quasicrystals. Philosophical Magazine A. 1996;**74**:697-708

[51] Ritsch S. Highly perfect decagonal Al–Co–Ni quasicrystals. Philosophical Magazine Letters. 1996;**74**:99-106

[52] Steurer W, Kuo KH. Five-dimensional structure analysis of decagonal $Al_{65}Cu_{20}Co_{15}$. Acta Crystallographica Section B. 1990;**46**:703-712

[53] Burkov SE. Structure model of the Al–Cu–Co decagonal quasicrystal. Physical Review Letters. 1991;**67**:614-617

[54] Yamamoto A. Science Reports of the Research Institutes, Tohoku University, Series A. 1996;**42**:207-212

[55] Saitoh K, Tsuda K, Tanaka M. Structural models for decagonal quasicrystals with pentagonal atom-cluster columns. Philosophical Magazine A. 1997;**76**:135-150

[56] Cockayne E, Widom M. Ternary model of an Al–Cu–Co decagonal quasicrystal. Physical Review Letters. 1998;**81**:598-601

[57] Deloudi S, Fleischer F, Steurer W. Unifying cluster-based structure models of decagonal Al–Co–Ni, Al–Co–Cu and Al–Fe–Ni. Acta Crystallographica Section B. 2011;**67**:1-17

[58] Wan W, Sun JL, Hovmöller S, Zou XD. RED data collection and data processing. 2013. Available from: http://www.calidris-em.com

[59] Sheldrick GM. In: Fortier S, editor. Direct Methods for Solving Macromolecular Structures. Dordrecht: Kluwer Academic Publishers; 1998. pp. 401-411

[60] Sheldrick GM. A short history of *SHELX*. Acta Crystallographica Section A. 2008;**64**:112-122

[61] Christensen J, Oleynikov P, Hovmöller S, Zou XD. Solving approximant structures using a "strong reflections" approach. Ferroelectrics. 2004;**305**:273-277

[62] Hovmöller S. *CRISP*: Crystallographic image processing on a personal computer. Ultramicroscopy. 1992;**41**:121-135

Chapter 5

Micro-/Nano-Structuring in Stainless Steels by Metal Forming and Materials Processing

Tatsuhiko Aizawa, Tomomi Shiratori and Takafumi Komatsu

Abstract

Austenitic stainless steel type AISI304 sheets and plates as well as fine-grained type AISI316 (FGSS316) substrates and wires were employed as a work material in the intense rolling, the piercing and the plasma nitriding. AISI304 sheet after intense rolling had textured microstructure in the rolling direction. Crystallographic state changed itself to have distorted polycrystalline state along the shearing plane by piercing, with the strain induced phase transformation. FGSS316 substrates were plasma nitrided at 623 K for 14.4 ks to have two-phase fine nanostructure with the average grain size of 100 nm as a surface layer with the thickness of 30 μm. FGSS316 wires were also plasma nitrided at the same conditions to form the nitrided surface down to the depth of 30 μm. This nitrided wire was further uniaxially loaded in tensile to attain more homogeneously nitrided surface nano-structure and to form the austenitic and martensitic fiber structure aligned in the tensile direction. Each crystallographic structure intrinsic to metals and metallic alloys was tailored to have preferable micro-/nano-structured cells by metal forming and nitrogen supersaturation. The crystallographic change by metal forming in a priori and posterior to nitriding was discussed to find out a new way for materials design.

Keywords: micro-/nano-texturing, stainless steels, rolling, piercing, plasma nitriding, a priori metal forming to nitriding, posterior nitriding to metal forming

1. Introduction

Most of metals and metallic alloys have crystalline structure, intrinsic to each material property. This crystalline structure is classified by several items; e.g., single and poly-crystals, grain size, grain boundary characteristics, crystallographic orientation, and so forth [1]. These items are controllable by mechanical and chemical interaction with internally and externally straining [2]. For examples, a single crystal changes itself to polycrystalline state by introduction of dislocations with sufficiently high density [3]. The original grains are much refined by intense rolling [4] and by high shear straining [5]. The initial grain boundaries are also tunable by shuffling process through their interaction with dislocations [6]. That is, the crystalline structure is tailored by metal forming and materials processing to have preferable grain size, crystallographic orientations, and grain boundaries [7–9].

Let us consider how to control the crystallographic structure of metals and metallic alloys by the technology of plasticity in metal forming and surface treatment. **Figure 1** depicts three case studies on the crystallographic structure change by rolling, shearing, and nitrogen supersaturation. Even when the initial grains are equiaxial, some of them are forced to partially align along the rolling direction. The skew distortion by intense rolling drives to shear the grains with spin-rotation as shown in **Figure 1a**. In case of embossing and piercing the sheet metals, the grains are distorted to plastically flow along the shearing plane with grain size refinement as depicted in **Figure 1b**. In the low temperature plasma nitriding, the plastic straining is induced by the nitrogen supersaturation into metallic lattices to form the slip-line system as shown in **Figure 1c**. Without externally applied stresses, the slip-line network is formed from the surface to the depth together with the nitrogen interstitial diffusion. In particular, the low temperature plasma nitriding process [10–15] works as a powerful means to demonstrate that austenitic and martensitic stainless steel substrates are hardened and modified to have two-phase structure with the average grain size of 0.1 μm. These previous studies proved that grain size as well as crystallographic structure should be significantly controlled by the materials processing other than the shearing process in metal forming.

In the present chapter, the crystallographic structure evolution of stainless steels during the rolling, the piercing, and the plasma nitriding at 623 K for 14.4 ks is first described to deduce the mechanism of microstructure evolution during metal forming and materials processing. Next, the uniaxial loading test of plasma nitrided work at 623 K is performed to investigate the possibility of further microstructure evolution during posterior metal forming. Through these experiments, the effect of the interstitial element concentration as well as the plastic straining on the crystallographic evolution is discussed to search for the materials science model to describe the interaction between the interstitial mobility and the plastic straining. In the following, EBSD (Electron Back Scattering Diffraction) is employed to make crystallographic analyses. This technique is based on the automatic analysis of the Kikuchi pattern by the excitation of the electron beam on the surface of the sample in SEM (Scanning Electron Microscope) [16]. Among several analytical tools in EBSD, the crystallographic orientation for each grain is described by IPF (Inverse Pole Figure) and the strain induced phase transformation is also analyzed by phase mapping. In addition, the equivalent plastic strain distribution is estimated by KAM (Kernel Angle Misorientation) mapping.

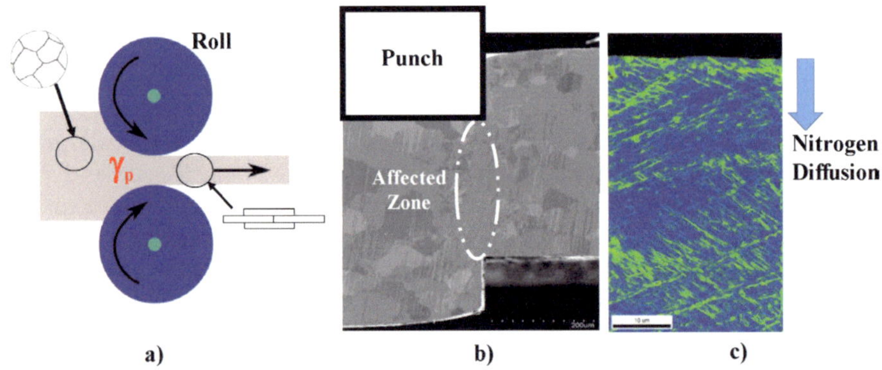

Figure 1.
Three types of crystallographic evolution during metal forming and plasma nitriding. (a) Intense rolling, (b) embossing and piercing, and (c) low temperature plasma nitriding.

2. Texture formation by intense rolling

Intense rolling with heat treatment has been utilized to fabricate the fine-grained stainless steel plates and sheets [4, 7, 17]. In the following, AISI304 sheet was employed as a work to reduce its thickness from 10 mm down to 1 mm by intense rolling. EBSD analysis was used to describe the crystallographic change in the rolled AISI304 sheet.

2.1 Rolling procedure

A typical rolling system was illustrated in **Figure 2a**. AISI304 sheet with the initial thickness of 10 mm was compressed and sheared between two work rolls in a single reduction. Since the reduction of thickness was 10% in this single rolling, nine steps were utilized to reduce the thickness down to 1 mm through this intense rolling. **Figure 2b** shows the rolled sheet by 90% reduction in thickness. This rolling is effective to reduce the average grain size of stainless steel works for embossing and piercing to be discussed in later.

2.2 Microstructure evolution by intense rolling

The original AISI304 sheet was characterized by three high intensity peaks in XRD analysis. As depicted in **Figure 3**, three peaks were detected in correspondence to γ (111), γ (200), and γ (220) planes in the austenitic phase. This microstructure changes to nearly full-martensitic phase; as also depicted in **Figure 3**, three martensitic peaks were detected as α’ (211), α’ (200), and α’ (110) besides for γ (220) and γ (111). This proves that original austenitic grains massively transform to martensitic ones in AISI304 sheets during the intense rolling. These martensitic grains can be inversely transformed back to austenite by heat treatment. This technique is useful to reduce the average grain sizes to be stated later.

2.3 Controlled crystallographic structure by intense rolling

In addition to the strain-induced phase transformation in **Figure 3**, the intense rolling has much influence on the crystallographic structure in AISI304. **Figure 4** depicts the inverse pole figure mapping, the KAM distribution, and the phase mapping on the cross-section of rolled AISI304 sheet, analyzed by EBSD. As stated before, the average grain size is reduced down to 1 μm in most of AISI304, since the plastic strains are applied to these regions as shown in **Figure 4a** and **b**. Low plastic strained regions in **Figure 4b** corresponds to assembly of larger grains laterally aligned in the rolling direction. Through comparison between **Figure 4b** and **c**, these

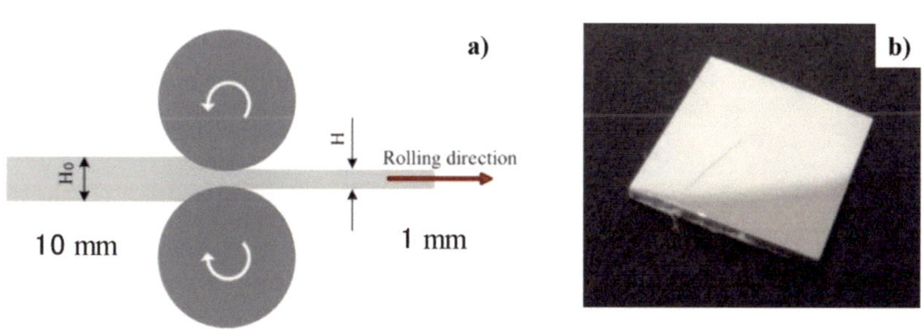

Figure 2.
Intense rolling process with high reduction in thickness. (a) Illustration on the rolling process in multi-steps for reduction of thickness by 90% and (b) rolled AISI304 sheet.

textures are classified into two zones. One is thin and long full-martensitic textures. The shorter and dotted textures are corresponding to the retained austenitic textures.

These textured crystallographic structures are further controlled by the low temperature plasma nitriding as well as the heat treatment to be discussed in later.

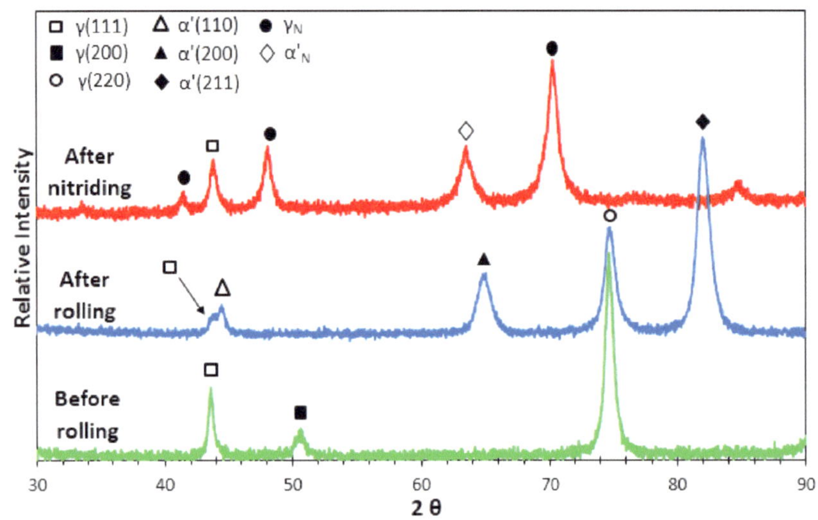

Figure 3.
Variation of XRD diagrams from the original AISI304 sheet before rolling to AISI304 after intense rolling and furthermore to rolled AISI304 sheet after low temperature plasma nitriding.

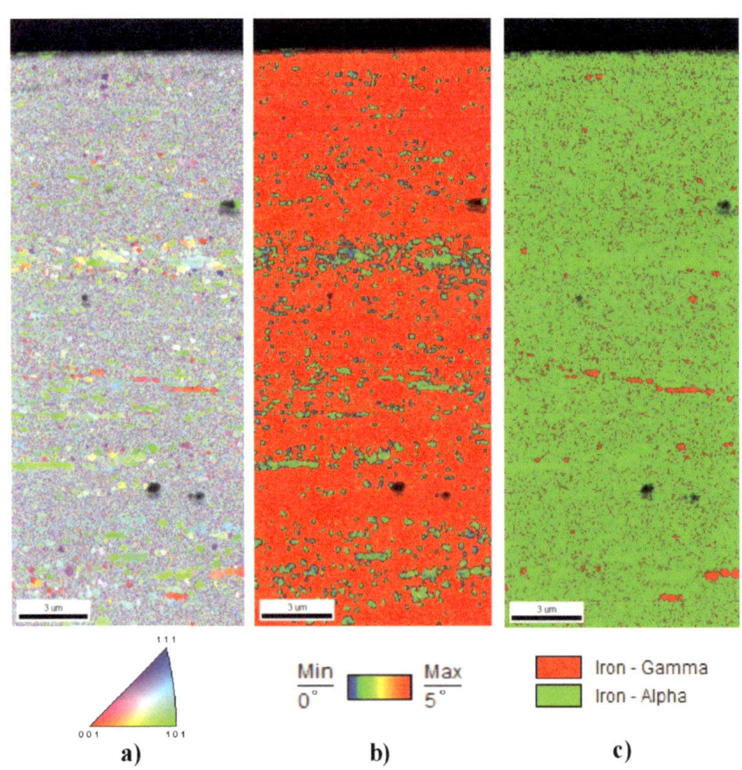

Figure 4.
EBSD analysis on the cross-section of rolled AISI304 sheet. (a) IPF mapping, (b) KAM distribution, and (c) phase mapping.

3. Crystallographic change by piercing

Fine piercing is an essential process in metal forming for accurate drilling of holes and for fine blanking. In the last decade, metastable austenitic stainless steel type AISI304 with fine grains has been developed by [18–20]; the effects of fine crystallographic structure on the elasto-plastic deformation have been closely studied in [21–23]. **Figure 5** shows three metastable austenitic stainless steel AISI304 sheets with the thickness of 100 μm, where the grain size was reduced by rolling process from the normal-grained sheet with the average grain size (D) of 7.5 μm. Two fine-grained AISI304 sheets were yielded to have D = 3.0 and 1.5 μm, respectively, by reverse transformation of the strain-induced martensitic phase. In this thermo-mechanical treatment, near-fully martensitic grains in **Figure 4c** are inversely transformed together with reduction of grain sizes [4, 7].

Their IPF maps analyzed by EBSD were shown on the cross section in the sheet width direction. When D = 7.5 μm, most of grains have a preferred orientation to [111] direction, as shown in **Figure 5a**. The crystallographic structure for D = 3 and 1.5 μm became nearly the same as D = 7.5 μm as depicted in **Figure 5b** and **c**.

3.1 Microstructure evolution by piercing

These AISI304 sheets with different grain sizes were pierced by CNC stamper under the same conditions; e.g., the punch diameter was 100 μm, the die diameter, 110 μm, and the clearance, 5% of sheet thickness. **Figure 6** compares IPF mapping as well as phase mapping on the cross-section of punched hole among three AISI304 sheets with the different grain size by D = 1.5, 3.0, and 7.5 μm. As commonly seen in **Figure 6a–c**, most of grains along the side surface of hole are distorted and refined in size. The shearing of AISI304 sheet by piercing leaves the process-affected zones along the shearing plane. The phase mapping in **Figure 5d–f** reveals that these process-affected zones are just corresponding to the martensitic phase [24]. That is, the original austenitic matrix to AISI304 sheets is forced to transform to martensite by the shearing strain during the piercing process. This strain-induced martensitic transformation is a non-diffusive shear transformation; each grain in the affected-process zone massively makes transformation [25, 26].

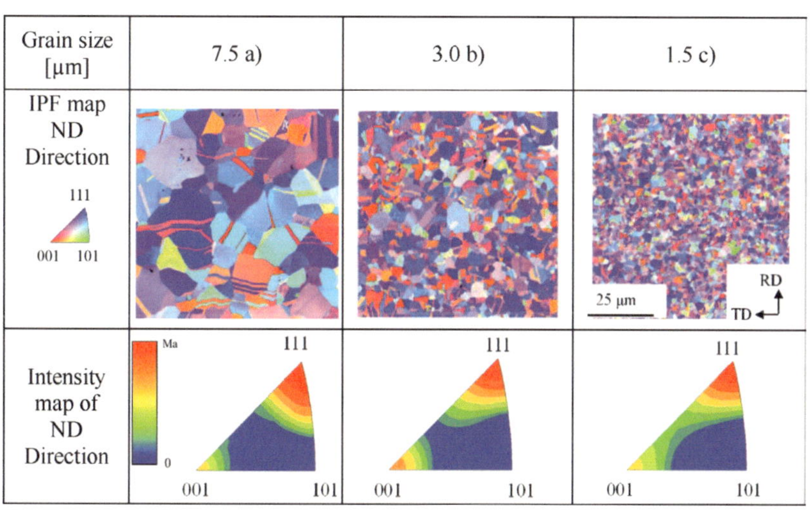

Figure 5.
IPF map and intensity map in the ND direction analyzed by EBSD. (a) Normal-grained AISI304 sheet with F = 7.5 μm, (b) fine-grained AISI304 sheet with D = 3.0 μm, and (c) fine-grained sheet with D = 1.5 μm.

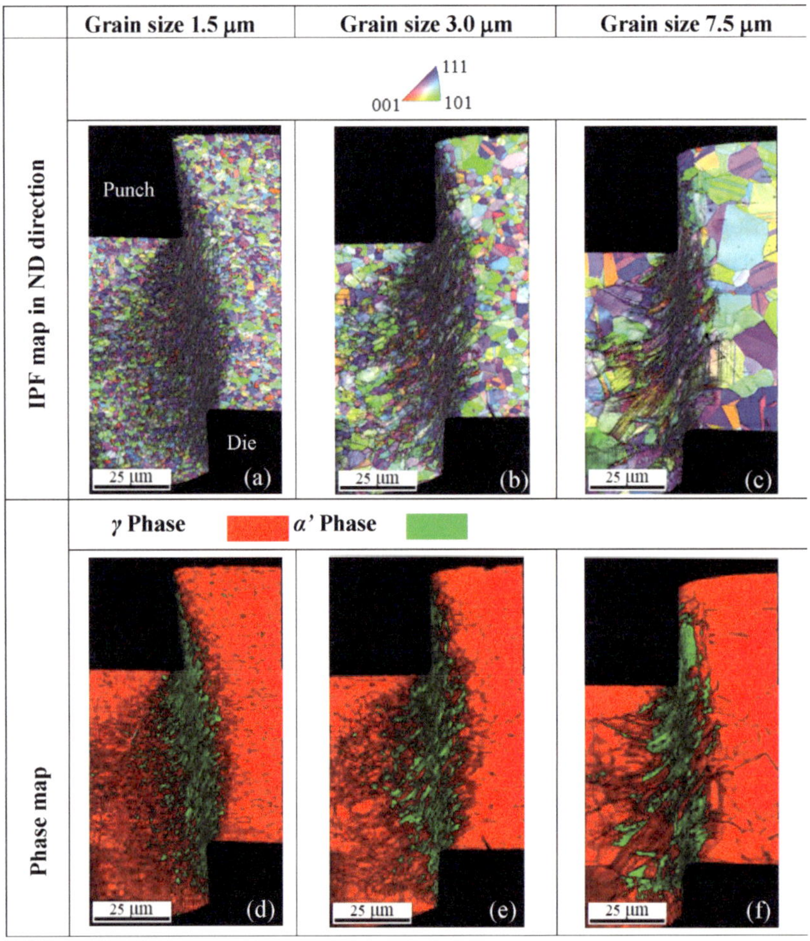

Figure 6.
EBSD analyses of IPF mapping in the ND direction and phase mapping on the cross-section of punched AISI304 sheet. (a) IPF map at D = 1.5 μm, (b) IPF map at D = 3 μm, (c) IPF map at D = 7.5 μm, (d) phase map at D = 1.5 μm, (e) phase map at D = 3 μm, and (f) phase map at D = 7.5 μm.

Let us consider the effect of average grain size on this formation of affected-process zones with phase transformation. When D = 7.5 μm, the zone boundary between the austenitic and martensitic phases is shaped to be irregularly jagged in **Figure 6d**. With decreasing the grain size, this phase boundary gradually is homogenized to be smooth as seen in **Figure 6e** and **f**. To be noticed, the volume of strain-induced martensitic phase in **Figure 6f** becomes larger than that in **Figure 6e**. This increase of strain-induced martensitic phase volume ratio causes the variation of the fracture length of the hole at the punch stroke direction [24].

3.2 Controlled crystallographic structure by piercing

The strain induced phase transformation in the process-affected zones by piercing is commonly observed in the metastable austenitic stainless steel sheets. Among them, the chemical components in each class of stainless steels have influence on the microstructure change by the piercing process. Two types of metastable austenitic stainless steel sheets were prepared as a work sheet with the thickness of 100 μm for piercing experiments under the same conditions as shown in **Figure 6**.

Figure 7 compares the phase mapping on the cross-section of punched holes between the stainless steel AISI316L and AISI304. When punching out the AISI316L sheet, the strain-induced martensitic phase hardly occurred along the shearing plane; no martensitic phase was present in **Figure 7a**. Instead of martensitic phase maps, the ferritic phase was strain-induced even by this piercing process as predicted by [27]. In fact, much ferrite was detected at the vicinity of die corner in **Figure 7a**. Let us consider why no martensitic phase but ferritic phase is induced by piercing the AISI316L sheet. As stated in [27], the nominal strain to induce the martensitic phase into AISI316L is two times larger than that to induce the ferrite phase. Hence, when piercing AISI3016L sheet, the ferritic phase is much easier to be induced than the martensitic phase. Therefore, even when the ferrite phase is induced along the shearing plane by the piercing process, each grain deformation is not hindered by shearing. As a result, the process-affected zone area in AISI316L becomes larger along the shearing plane in **Figure 7a** than that in AISI304 in **Figure 7b** [28].

The difference of crystallographic structure between pierced AISI304 and AISI316L sheets reveals that the process-affected zone formation could be controlled by the strain-induced phase transformation during the piercing process. EBSD analysis is really a well-defined means to describe the relationship between the strain-induced phase transformation by piercing and the crystallographic structure of pierced stainless steel sheets.

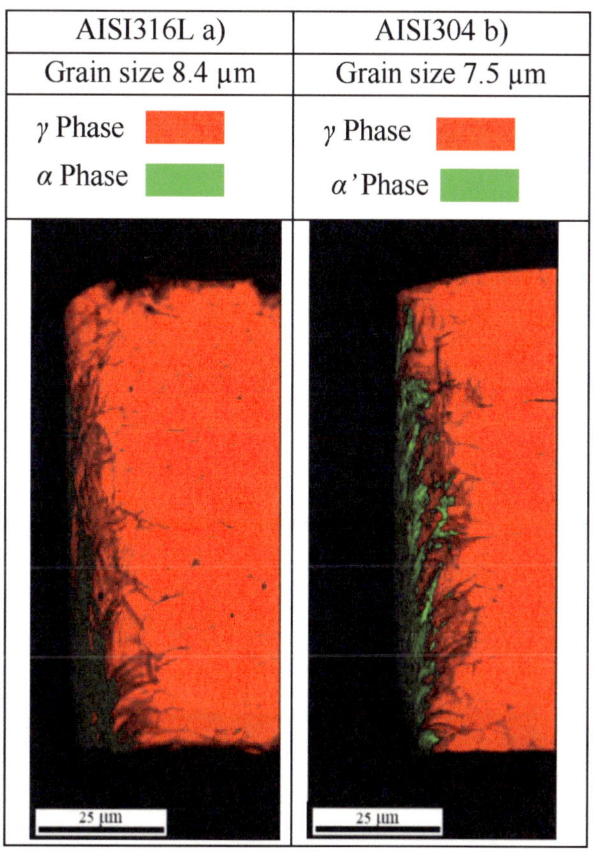

Figure 7.
Comparison of phase mapping on the cross-section of fully pierced AISI316L and AISI304 austenitic stainless steel sheets. (a) Punched-out AISI316L sheet and (b) punched-out AISI304 sheet.

4. Nanostructuring by low temperature plasma nitriding

Lower temperature plasma nitriding process of austenitic stainless steels than 700 K is governed by the nitrogen supersaturation with nitrogen interstitial occupation of octahedral vacancy sites of fcc-structured supercells as well as the nitrogen diffusion through refined grain boundaries and slipping lines [13–15]. FGSS316 plates and wires are employed to describe the nano-structuring process with grain size refinement by this plasma nitriding.

4.1 Low temperature plasma nitriding

High density RF (Radio Frequency) – DC (Direct Current) discharging plasma nitriding system is utilized to generate the nitrogen-hydrogen plasmas. **Figure 8** illustrates a typical hollow cathode device for homogeneously nitriding a single FGSS316 wire. RF-nitrogen/hydrogen plasma is ignited to surround the wire surface by a cylindrical plasma sheath. The activated nitrogen atoms (N^*) and ions (N^+) as well as the NH radicals are enriched in this sheath to increase the nitrogen ion density up to 1×10^{18} ions/m^3. Under this plasma processing condition, the nitrogen solute diffuses into the depth of FGSS316 matrix in wire to form the nitrided layer.

The inner nitriding process with nitrogen supersaturation is described by the multi-dimensional relation in **Figure 9**. When some nitrogen interstitial atoms diffuse and supersaturate the fcc-structured lattices in **Figure 9a**, their original lattice constant (Λ_0) increases to Λ by occupation of nitrogen interstitials into the octahedral vacancy sites in them. Other nitrogen atoms diffuse to further depth through the grain boundaries. Then, the nitrogen supersaturated (NS) zone expands with the elastic strain (ε_e) while unsaturated (US) zone does not deform; the strain incompatibility occurs on the boundary between NS and US zones. The misfit distortion (β_{misfit}) is induced along this zone boundary as depicted in **Figure 9b**; e.g., $\beta_{misfit} = \omega_{spin} + \varepsilon_p$, where ω_{spin} is a spin tensor to rotate the NS zone and to generate the crystallographic misorientation into a current zone, and ε_p, a plastic strain tensor to compensate for the strain incompatibility across the zone boundary. Nitrogen solutes further diffuse into the depth of grains through these zone boundaries, as shown in **Figure 9c**.

Through this multi-dimensional inner nitriding, high elastic strain energy density in NS zones drives the phase transformation from austenite to martensite. The spin rotation of zones and sub-grains advances with nitrogen diffusion and

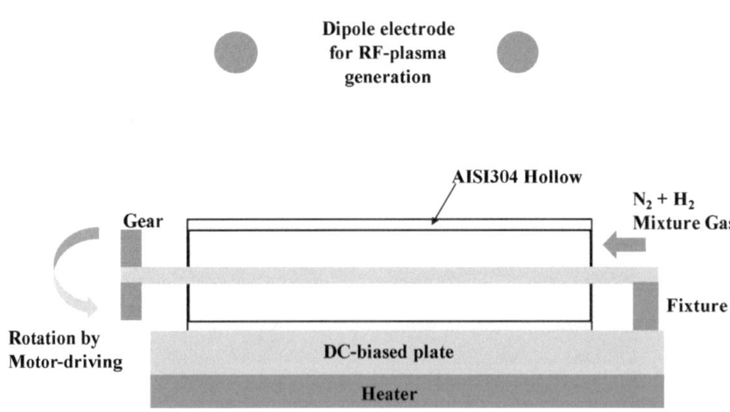

Figure 8.
Experimental setup for low temperature plasma nitriding of steel wires with use of the hollow cathode device. The same DC-bias was applied to the hollow and the FGSS316 wire.

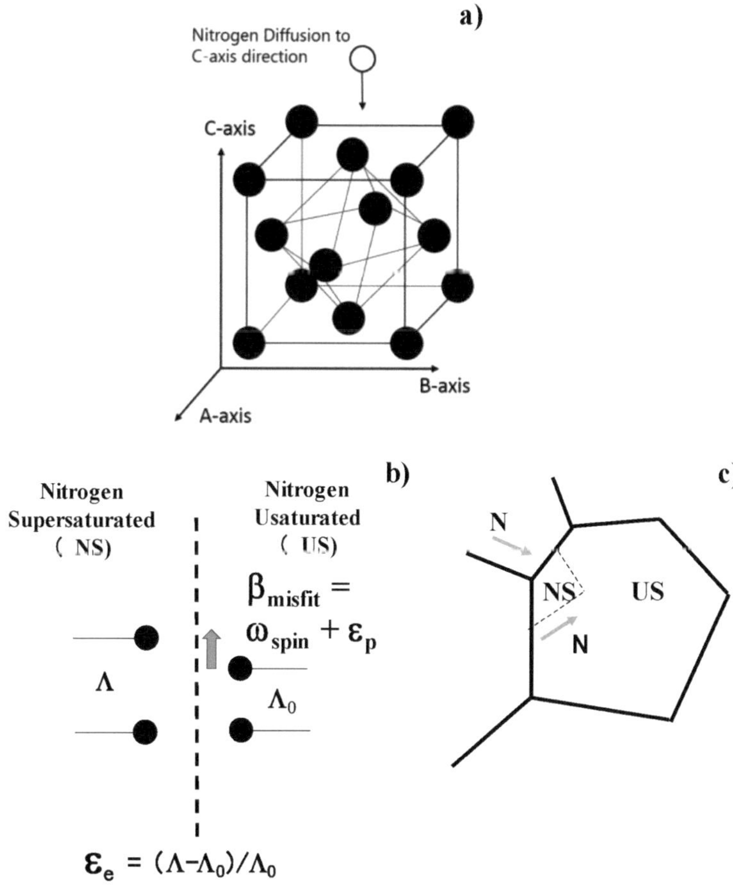

Figure 9.
Multi-dimensional relation in the nitrogen interstitial atom diffusion and supersaturation in the low temperature plasma nitriding.

supersaturation to refine the crystallographic structure. NS zones accompany with the plastic strain distribution along the NS-US zone boundaries.

This theoretical model is experimentally demonstrated in the following. As stated in [13–15], the inner nitriding advances homogeneously from the surface to the nitriding front end; it is rather difficult to experimentally describe each fundamental process separately from other processes in **Figure 9**. A normal-grained AISI316 plate is employed as a work material and nitrided at 623 K for 14.4 ks to decelerate the nitrogen diffusion rate and to describe the synergetic relation among the nitrogen supersaturation, the phase transformation, the plastic straining, the grain size refinement and the local nitrogen diffusion.

4.2 Microstructure evolution by inner nitriding in depth

Figure 10 depicts the homogeneous and heterogeneous inner nitriding processes in the nitrided AISI316 at 623 K for 14.4 ks. Under this nitriding condition, the high nitrogen solute content, [N], around 5 mass% is present down to the nitriding front end (NFE) at the depth of 30 μm. To be noted, [N] remains to be 1 mass % even below NFE, as shown in **Figure 10a**. This implies that homogeneous nitriding advances to NFE and changes to be heterogeneous by localization in nitrogen diffusion below NFE. In fact, the phase mapping as well as the crystallographic structure changes drastically across this NFE, which was indicated by the gray two-dots chain line in **Figure 10**.

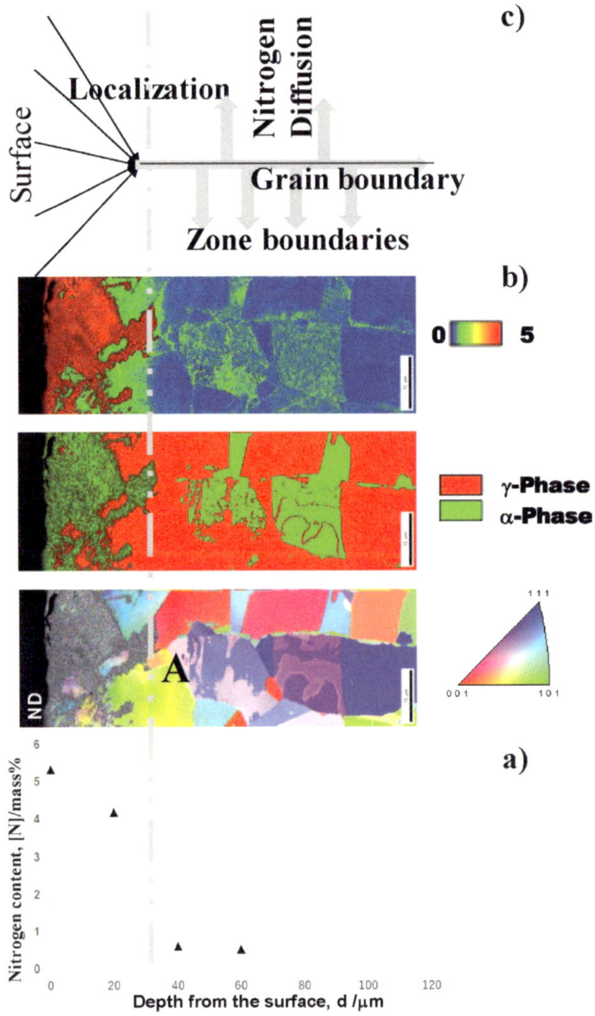

Figure 10.
Experimental demonstration on the inner nitriding process. (a) Nitrogen content depth profile, (b) IPF mapping, phase mapping and KAM distribution, and (c) inner nitriding mechanism.

Due to EBSD analysis, the synergetic relation is described by the inverse pole figure, the phase mapping and the plastic strain distribution below NFE in **Figure 10b**. The coarse grains near the grain boundary were only nitrided to change their microstructure; e.g., the austenitic and martensitic zones exclusively distribute in an A-grain. The retained austenitic zones in this grain had higher plastic strains (or higher KAM) and different crystallographic orientations from the martensitic zones, the plastic strains of which were much lower than these austenitic ones. This formation of γ- and α'-zone mixture with plastic straining and crystallographic rotation just corresponds to the multi-dimensional inner nitriding process. As depicted in **Figure 10c**, the nitrogen diffusion localizes across NFE so that the nitrogen selectively diffuses along the original grain boundary below NFE. This nitrogen main stream branches into the neighboring grains to this gran boundary. In each grain, the nitrogen further diffuses through the zone boundaries to drive the phase transformation, the plastic straining, and the refinement of zone sizes. Next, more precise EBSD analysis is made on this A-grain.

As compared between **Figure 11a** and **b**, little plastic strains were detected in the α'-zone of A-grain while the surrounding γ-zones were much plastically strained.

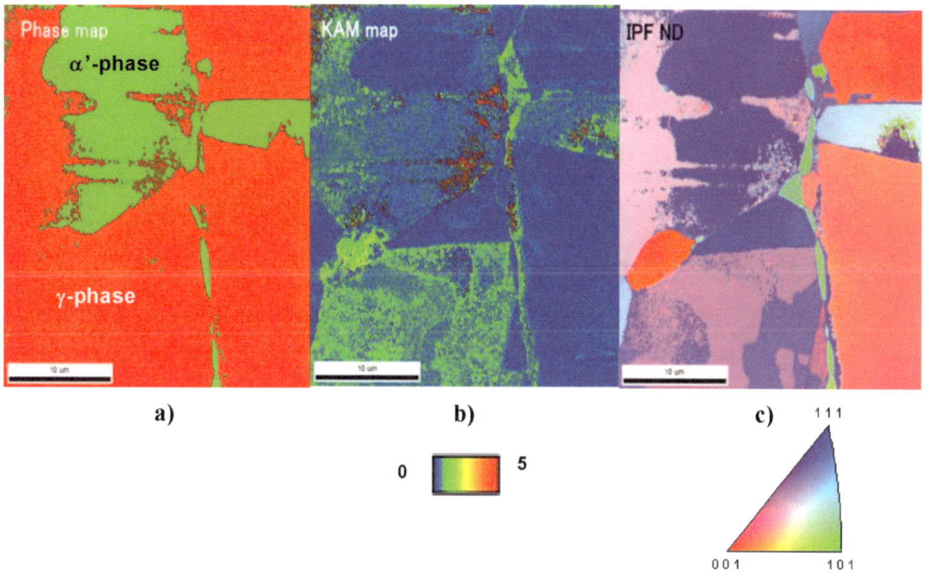

Figure 11.
Synergetic relation in the inner nitriding at 623 K, locally observed in the grain-A. (a) Phase mapping, (b) KAM distribution, and (c) IPF mapping.

This proves that some of NS-zones massively transform to α'-zones since the elastic strain energy density reaches to the critical level and that plastic strains are induced only in the neighboring γ-zones to α'-zones to compensate for misfit strain between NS-zones and surrounding unsaturated γ-zones. Comparing **Figure 11b** and **c**, these highly strained zones are just corresponding to crystallographically refined zones. This assures that spin rotations are induced together with the plastic strains by misfit distortion on the distributed zone boundaries to refine each zone size. Owing to this synergetic mechanism, the refined zones have mutual boundaries with high misorientation angles. This grain size refinement reflects on the high strength and hardness of the nitrided layer by the low temperature plasma nitriding.

4.3 Microstructure evolution in FGSS316 wires

As had been discussed in [29, 30], the initial grain size of AISI316 has influence on the inner nitriding behavior; homogeneity in nitriding process is enhanced with reduction of initial grain size. This finding suggested that fine-grained AISI316 or FGSS316 structural components and parts could be homogeneously nitrided on the surface with a little effect of heterogeneous nitriding on the microstructure below NFE. In the following, a FGSS316 wire is nitrided at 623 K for 14.4 ks and its microstructure is precisely analyzed to describe the formation of nitrided layer as well as the microstructure evolution at the surface and below NFE during this low temperature nitriding. **Figure 12** depicts the nitrogen solute mapping, phase mapping, and crystallographic structure on the lateral and longitudinal cross-sections of nitrided FGSS316 wire, respectively. The wire surface is surrounded by two-phase fine-grained layer with the thickness of 30 μm and the average nitrogen content of 5 mass% as seen in **Figure 12a** and **d**. The retained austenitic phase grains with larger size than nano-structured grains are present in the inside of nitrided layer as seen in **Figure 12b**, **c**, **e**, and **f**, respectively. This reveals that inner nitriding process advances almost homogeneously but that its synergetic relation is retarded in some parts to leave the nitrogen supersaturated γ-grains without phase

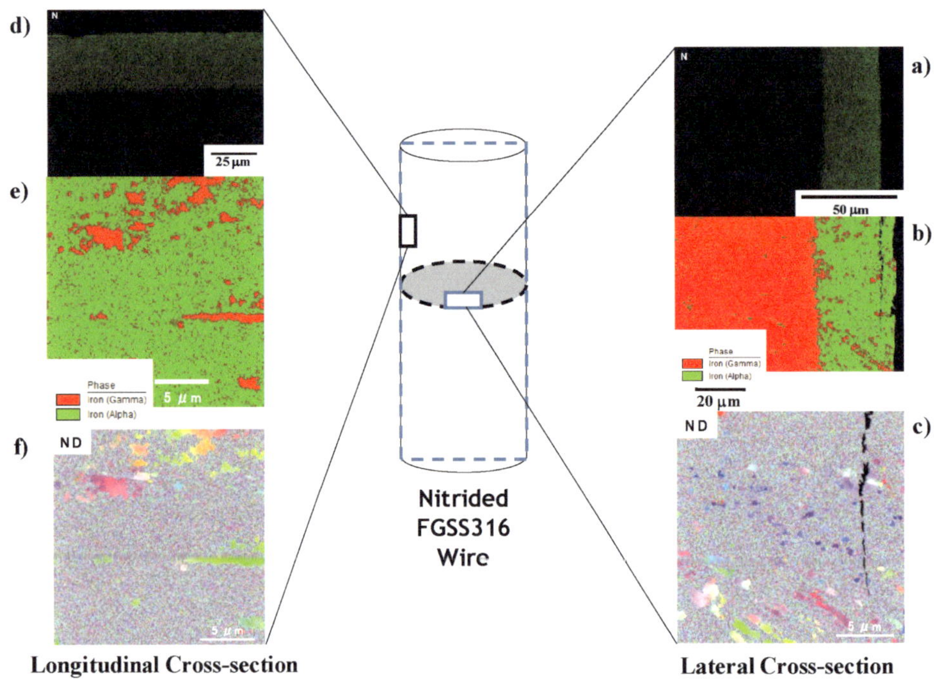

Figure 12.
Microstructure and nitrogen mapping across NFE on the lateral and longitudinal cross-sections of nitrided FGSS316 wire at 623 K for 14.4 ks. (a) Nitrogen mapping, (b) phase mapping, and (c) IPF mapping in the nitrided layer, on the lateral cross-section. (d) Nitrogen mapping (e) phase mapping in the nitrided layer, and (f) IPF mapping in the nitrided layer, on the longitudinal cross-section.

transformation and grain size refinement. In the following section, these retained γ-zones are employed as a marker to investigate the effect of uniaxial loading on the nitrided surface layer.

Let us analyze the effect of inner nitriding on the FGSS316 matrix below NFE. **Figure 13a** shows the initial microstructure of FGSS316 wire. This microstructure of FGSS316 before nitriding has equiaxial crystallographic structure with the average grain size of 2 μm. **Figure 13b** and **c** shows the crystallographic microstructure after nitriding on the lateral and longitudinal cross-sections of wire, respectively. The original microstructure is modified to form the skewed linear zones with finer grains as pointed by "b" in **Figure 13c**. These nitrogen-supersaturated zones consist of the transformed α'-zones and their surrounding plastic-strained. Just as discussed in **Figures 10** and **11**, the inner nitriding process advanced heterogeneously in the depth below NFE even when nitriding the FGSS316 wire.

4.4 Controlled crystallographic structure by plasma nitriding

Different from the metal forming, the plastic strains are induced along the network of subgrain and zone boundaries, which are newly generated in the inside of original matrix by nitrogen diffusion and supersaturation. Since this network also works as new nitrogen diffusion paths, further nitrogen diffusion and supersaturation advances into the depth of original grains. This concurrent co-working of nitrogen diffusion and supersaturation sustains the synergetic relation among the phase transformation, the plastic straining, and the grain size refinement. In particular, fine zone network results in refined nano-sized crystallographic structure after plasma nitriding.

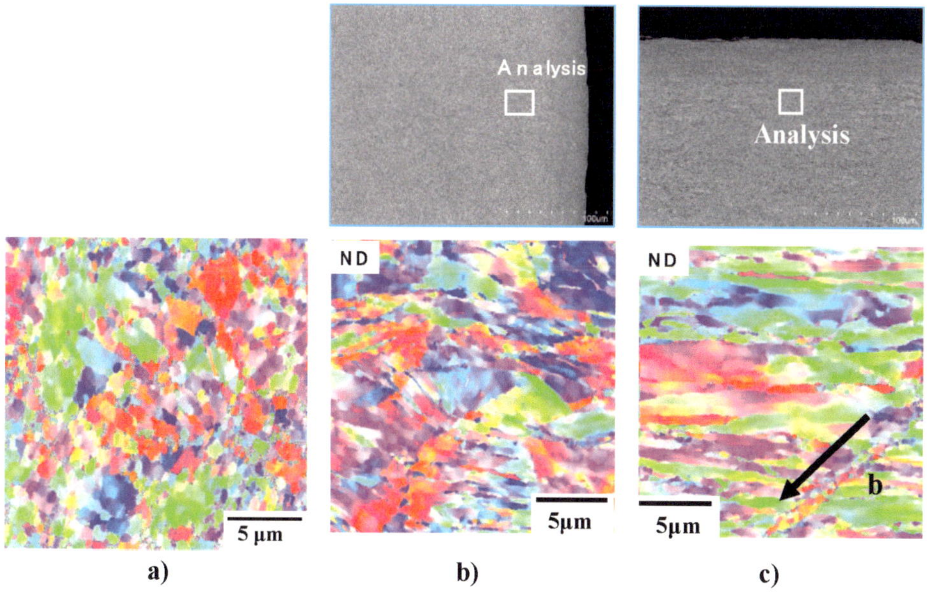

Figure 13.
Comparison of the crystallographic structure before and after plasma nitriding. (a) Initial microstructure of FGSS316 wire, (b) microstructure of FGSS316 matrix on its lateral cross-section, and (c) microstructure of FGSS316 matrix on its longitudinal cross-section.

5. Formation of bundle structures in nitrided wire by uniaxial loading

The nitrided FGSS316 wire is further uniaxially loaded in tensile. The austenitic grains are continuously linked between its nitrided layer and inner matrix. This nitrided layer had influence on the mechanical response of wire since the same elasto-plastic strains are applied to these two regions. Precise microstructure analysis is also made to describe the microstructure evolution of this nitrided wire during the uniaxial loading.

5.1 Uniaxial loading procedure

The uniaxial tensile loading test was performed by using the precision universal testing machine AUTOGRAPH AGS-X 10 kN (SHIMADZU Co. Ltd.). This uniaxial loading was terminated when the maximum applied load reached 6 kN before fatal ductile fracture for microstructure analyses. The applied load and stroke were in situ monitored by the load cell and linear scaler, respectively. The bare FGSS316 wire without nitrided layer has an ultimate strength (σ_U) of 1.18 GPa at the stroke (δ) of 4.9 mm, or, at the nominal strain (ε) of 0.17.

5.2 Microstructure evolution during uniaxial loading at RT

A normal FGSS316 wire elongates in the tensile direction at room temperature (RT) without significant change of microstructure except for the formation and coalescence of voids near the fatal ductile fracture of wire [31]. Microstructure of high carbon steel wire is sensitive to drawing process since it has a multi-dimensional microstructure, where each austenitic grain boundary houses pearlite blocks and each block consists of pearlite colony with the same lamellar structure of cementite (or θ-phase) and lamellar ferrite [32]. However, little microstructure change occurs at the RT in its uniaxial tensile loading. A composite wire also has no microstructure evolution before pop-out of fibers from core matrix [33]. Let us investigate the microstructure at the nitrided layer and in the matrix of wire, respectively, and describe the effect of posterior elasto-plastic straining to nitriding on the microstructure evolution in wire.

Figure 14 depicts the plastic strain distribution and phase mapping on the lateral and longitudinal cross-sections of nitrided FGSS316 wire, respectively, after uniaxial loading. The retained austenite regions with larger size than fine grained two-phase nitrided layer in **Figure 12** completely disappear at the nitrided layer. Every nitrided layer has homogeneous two-phase microstructure with fine grain sizes. Compared with the plastic strain distribution before uniaxial loading, the whole nitrided layer is uniformly subjected to high plastic straining. This reveals that the applied plastic strains by uniaxial loading drives the synergetic process of inner nitriding at the retained austenitic zones and induces the phase transformation from the retained γ-phase to γ–/α'-phase fine mixture.

The microstructure as well as the phase in the matrix below NFE is modified by this uniaxial loading. Nearly full austenitic phase of matrix before loading changes to mixture of γ-fibers and transformed α'-fibers. As shown in **Figure 14b** and **d**, these α'-fibers with its lateral size of 0.5 μm are aligned along the loading direction to form a bundle structure together with γ-fibers. The volume fraction of these α'-bundles increases from the vicinity of NFE to the depth in matrix. This suggests that the nitrided FGSS316 wire fractures of α'-bundles in ductile at its center when this fraction reaches to the critical maximum.

Let us compare the above phase mapping in matrix with the plastic strain distribution in **Figure 14**. The high plastic straining zones are just corresponding to the α'-fiber zones. This proves that this γ to α'-phase transformation during the uniaxial loading is induced by the high plastic straining of γ-zones. That is, the highly strained γ-grains are plastically strained and aligned along the loading direction to form the γ-fibers. During this microstructure evolution, some of γ-fibers transforms massively to α'-fibers.

The phase transformation and formation of bundle microstructure in the above reflects on the crystallographic structure after uniaxial loading. **Figure 15** depicts the IPF mapping at the nitrided layer and in the matrix on the lateral and

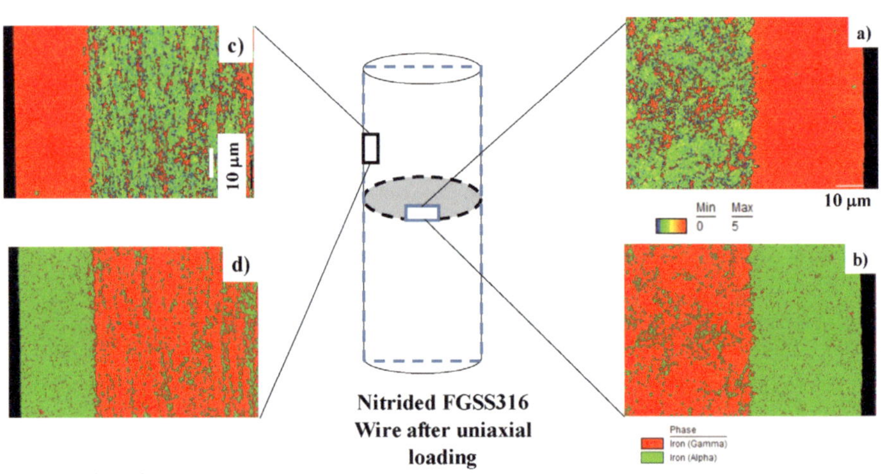

Figure 14.
Plastic straining and phase mapping on the lateral and longitudinal cross-sections of nitrided FGS316 wire after uniaxial loading. (a) Plastic strain lateral distribution, (b) phase mapping in lateral, (c) plastic strain longitudinal distribution, and (d) phase mapping in longitudinal.

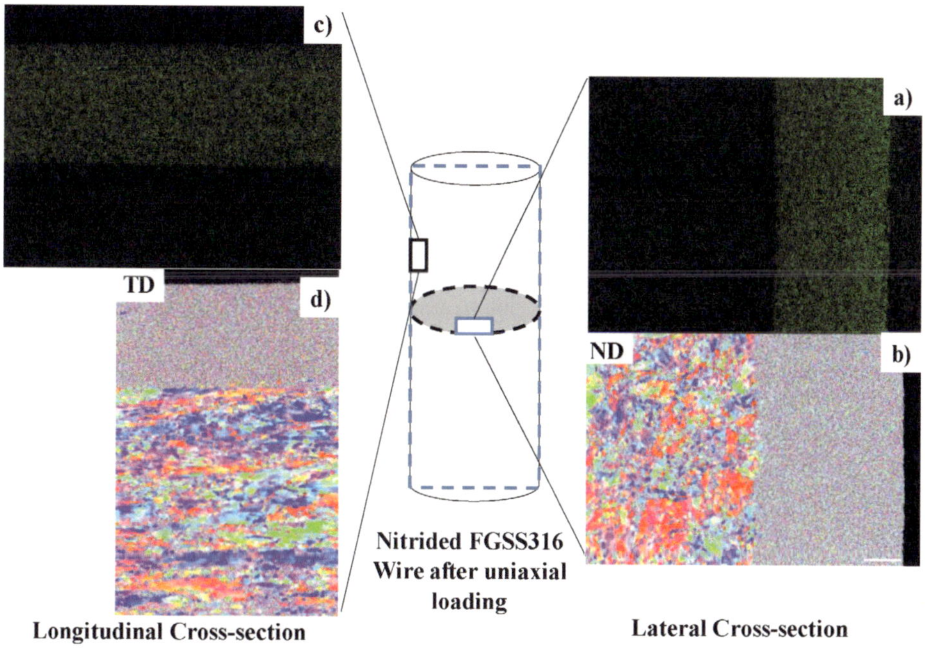

Figure 15.
Nitrogen mapping and crystallographic structure on the lateral and longitudinal cross-sections of nitrided FGSS316 wire after uniaxial loading. (a) Nitrogen mapping in lateral, (b) lateral IPF mapping in ND, (c) nitrogen mapping in longitudinal, and (d) longitudinal IPF mapping in TD.

longitudinal cross-sections of wire, respectively. The γ-phase matrix before uniaxial loading has crystallographic structure with the average grain size of 2 μm and without preferred orientations. After loading, these γ- and α'-grains are aligned along the loading direction respectively to form α'- and γ-bundles. In particular, most of α'-bundles have unique (111) directions. This might be because the transformed bundles are aligned in the tensile directions under the constrained conditions by the nitrided layer, surrounding the matrix. Each initial γ-grain is forced to deform elasto-plastically only in the tensile direction and to form γ-fibers and γ-bundles since no plastic deformation is allowed in the lateral direction under the confinement of nitrided layer. Some of strained γ-fibers makes massive transformation to α'-fibers; these α'-fibers are assembled into a single α'-bundle with the preferred orientation to (111). After [26, 34], the easiest crystallographic orientation to form this martensitic fibers and bundles is thought to be (111) tensile direction.

Figure 15 reveals that the nitrided layer surrounding the matrix has homogeneous super fine-grained two phase structure after uniaxial loading. No cracks and defects are seen on the nitrided layer surface and in the inside of layer; the fatal fracture of this nitrided FGSS316 wire occurs as the ductile fracture of matrix as explained before. The fine continuous interface between the nitrided layer and the matrix also suggests that elasto-plastic compatibility is preserved across this interface.

6. Discussion

The polycrystalline materials are generally described by the grain boundary characteristics and crystallographic orientation of each constituent grain as well as its grain size [35]. Each grain boundary energy is determined by the misfit orientation angle between adjacent grains. The compatible grain boundary has low energy enough to stack some amount of dislocations; while the incompatible one has high energy enough to interact with dislocations [36]. In the metal forming of these polycrystalline materials or in the nitrogen supersaturation process, the dislocations as well as the slipping lines and planes interact with their grains and grain boundaries. In intense rolling, the grains are sheared and deformed to align their crystallographic orientation along the rolling direction and to form the textured microstructure. The applied elastoplastic distortion by rolling works to induce the phase transformation during rolling by high elastic strain energy density, to shear and elongate the grains by the plastic strains, and to spin the grain orientations toward the preferred one along the rolling direction. As reported in [37], the grain size is reduced down to sub-microns by repetitive rolling; due to small misorientation angles between adjacent grains, they are easy to agglomerate into larger grains by heat treatment.

In piercing, the grains near the shearing plane are affected by the applied elasto-plastic distortion [38]. The increase of elastic strain energy density during piercing induces the phase transformation. The grains are sheared and fractured along the shearing plain. The original crystallographic structure of work materials is changed to align along the shearing plane by severe spinning with piercing. Although the grains are refined at the vicinity of shearing plain, the misorientation angles among them are small enough to be identified as nearly the same grain. Even in other metal forming processes than two in the above, their mechanical interactions of elasto-plastic distortion with crystallographic structure is described by the strain-induced phase transformation, the shear deformation, the grain size refinement as well as the elastic recovery from the elasto-plastically strained state in unloading.

On the other hand, no elastoplastic distortion was directly applied to granular structure by the low temperature plasma nitriding. Instead of this direct straining, a large elastic distortion is induced into the nitrogen supersaturated zones in the work materials by lattice expansion. This distortion reaches to 10%, enough to drive the phase transformation in zones as well as the plastic distortion to compensate for the misfit on the zone boundaries between the nitrogen supersaturated and unsaturated ones. The symmetric component of this distortion works as a shearing strain tensor to form new slipping lines network across the original grain boundary. The asymmetric one drives spin-rotation in each zone to form the zone boundaries with high misorientation angles and to significantly refine the original grain size of work materials. Since those newly built-up zone boundaries play as a nitrogen diffusion path, this process advances concurrently with the nitrogen solute diffusion from the surface to the depth of materials. Since the zone size ranges in the nanometer order, the stainless steel work materials are covered by two-phase, nano-grained, nitrogen steel surface layer. The smallest zone size is determined by the mechanical balancing between the elastic straining in each material supercell and the slip-line formation surrounding it in the nitrogen supersaturation [39]. Precise TEM analyses down to the atomic scale are useful to describe the nitrogen interstitial atom distribution in the supercell [40].

Even when the rolled and pierced steel specimens are uniaxially loaded, their microstructure never changes themselves at room temperature before fracture. Their uniaxial stress-strain curves are determined by the original microstructure of as-rolled and as-pierced materials since no mechanical interaction occurs between the controlled grains by previous metal forming and the applied strain by uniaxial loading. On the other hand, the in situ microstructure evolution takes place during the uniaxial loading of the nitrided FGSS316 wire. The nitrogen supersaturated layer works a double role under this uniaxial tensile loading. The original FGSS316 matrix inside the wire is elastically supported by this nitrided layer. Since the austenitic grains are continuously linked with those in the nitrided layer, the matrix inside the wire is intensely elongated to change its microstructure to two-phase fibrous grains by the uniaxial loading. The microstructure in the as-nitrided layer is also affected by uniaxially applied plastic strains. The retained austenitic grains with large size in the nitrided layer change themselves to two-phase, fine grains, which are the same as homogeneously nitrided nanostructure before uniaxial loading. This in situ refinement of granular structure in the nitrided layer reflects on the hardness profile. The original hardness of as-nitrided layer is 1400 HV; this is further enhanced up to 1600 HV after uniaxial loading. This proves that nitrogen could diffuse locally to the retained austenitic zones along the slip-lines by externally applied plastic straining and drive the nitrogen supersaturation process in them for refinement of their microstructure. Owing to the elastic constraint by the nitrided layer, the work-hardening process during the uniaxial loading is enhanced in the wire matrix to attain higher ultimate stress (σ_U); e.g., σ_U = 1.23 GPa in the nitrided FGSS316 wire at δ = 5.7 mm or at ε = 0.19.

This in situ microstructure evolution by uniaxial loading posterior to nitriding, suggests further possibility of crystallographic control to improve the mechanical properties of metallic works, takes place in the nitrided layer and in the matrix inside without mutual interactions. The microstructure evolution of low temperature nitrided members and parts in the above must be enhanced during warm and hot processing. After recent work on the high carbon steel wire during drawing [41, 42], significant reduction of lamellar ferrite distance as well as free carbon dissociation from the cementite lamellar structure in the perlite colony and block are responsible for high strengthening of high carbon steel wires. This implies that further carbon supersaturation is a key process to drive the in situ evolution to the preferred crystallographic microstructure to higher strength of wire. Owing to the equivalent role between carbon and nitrogen solutes to be working as an interstitial atom in steel [43], local nitrogen mobility from nitrogen supersaturated zone with high nitrogen content to NS-zones with lower nitrogen content could drive the in situ nitrogen alloying process of wire matrix during warm-/hot-drawing and rolling.

In this nitriding a priori to metal forming, FGSS316 wire was first nitrided and then uniaxially loaded. How about the plasma nitriding of the rolled AISI304 plate? As shown in **Figure 3**, the nearly full-martensitic phase of rolled AISI304 plate

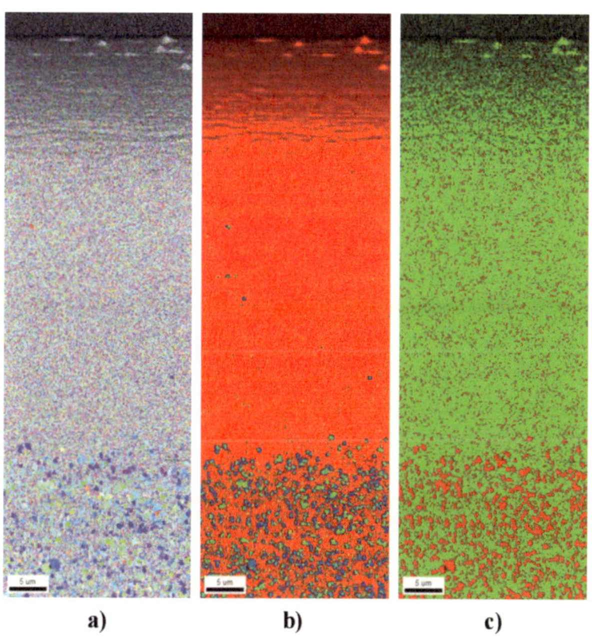

Figure 16.
Crystallographic structure on the cross-section of the rolled AISI304 plate after plasma nitriding at 673 K for 14.4 ks. (a) IPF mapping in the normal direction, (b) KAM distribution, and (c) phase mapping.

changed to a mixture of nitrided austenitic and martensitic phases. Since the original martensitic and austenitic peak positions shift to the low angle of 2θ and their peak widths become significantly broad, this mixture composes of the fine grained austenitic and martensitic zones with nitrogen supersaturation.

EBSD was also employed to describe this microstructure change of rolled AISI304 plate after nitriding. As shown in **Figure 16a**, the textured structure of rolled AISI304 completely disappeared and changed to fine-grained structure without preferred crystallographic orientation. This change is driven by high plastic straining in **Figure 16b**; every original grains with and without textures by rolling is plastically strained and spin-rotated by the nitrogen supersaturation to form homogeneous fine-grained structure. As depicted in **Figure 16c**, this fine microstructure consists of two phase with the fraction of martensite by 70%. This dramatic crystallographic structure evolution proves that posterior nitriding to metal forming is useful to further control the microstructure of stainless steels.

7. Conclusion

In the metal forming like intense rolling and fine piercing, the microstructure of work materials is changed by the applied plastic distortion with less influence to tool materials. In the rolling process, the original austenitic phase of stainless steels changes to be nearly full martensitic and to have textured microstructure with the preferred orientation to the rolling direction. This crystallographic structuring is intrinsic to the microstructure change by shearing with the reduction of thickness. In the piercing process by shear localization, the austenitic work material after piercing has new sheared and fractured surfaces including the affected zones. These zones consist of the phase-transformed martensite, the work-hardened austenite, and the elastically recovered zones. This crystallographic structure change is precisely described by EBSD on the cross-section of pierced work materials. In addition, various factors influence on this structure change including the grain size of work materials and the shear localization control as well as the chemical components in stainless steels.

In the low temperature plasma nitriding, no plastic strains are externally applied to work materials but nitrogen interstitial atoms are distributed from their surface to their depth with high concentration. Owing to the synergetic process in this inner nitriding process, the plastic distortion is concurrently induced by nitrogen diffusion and supersaturation. Since the nitrogen solute is homogeneously distributed in the nitrided layer with high content, the plastic distortion tensor also uniformly distributes in this layer. This homogeneous plastic distortion changes the normal crystallographic structure of AISI316 plates and wires; e.g., fine-grained AISI316 (FGSS316) microstructure of wires with the average grain size of 2 μm changes to the super-fine grained, two phase structure with the average grain size less than 0.1 μm. During this homogeneous nitriding, the retained austenitic zones distribute in the nitrided layer.

A priori nitriding to cold metal forming is a way to significantly control the microstructure and mechanical properties. The nitrided FGSS316 wire is elasto-plastically strained in the uniaxial direction so that the whole nitrided layers have fine-grained two-phase structure without retained austenite. This microstructure evaluation in local reflects on the homogeneous increase of hardness in the nitrided layer. This local interaction between nitrogen solute mobility and externally applied plastic strains at room temperature reveals that the microstructure and mechanical properties of nitrided work materials could be modified and improved by the metal forming posterior to the nitriding. In particular, the warm and hot post-treatment by drawing, rolling, forging, and stamping has capability to control the crystallographic structure of nitrided parts and components.

Posterior nitriding to metal forming is another way to refine the microstructure and to improve the mechanical properties. In case when the rolled AISI304 plate is further nitrided at 673 K, its textured microstructure completely changes to super-fine grained, two phase structure. The intense plastic straining by nitrogen supersaturation also plays a role to control the crystallographic structure.

Acknowledgements

The authors would like to express their gratitude to Mr. T. Inohara (LPS-Works, Co., Ltd.), Mr. T. Yoshino, and Y. Suzuki (Komatsu-Seiki Kosakusho, Co., Ltd.) for their help in experiments. This study was financially supported by the METI-Program on the Supporting Industries at 2019.

Conflict of interest

The authors declare no conflict of interest.

Author details

Tatsuhiko Aizawa[1]*, Tomomi Shiratori[2] and Takafumi Komatsu[3]

1 Surface Engineering Design Laboratory, SIT, Tokyo, Japan

2 University of Toyama, Toyama, Japan

3 Komatsu Seiki Kosakusho, Co., Ltd., Suwa, Nagano, Japan

*Address all correspondence to: taizawa@sic.shibaura-it.ac.jp

References

[1] Callister WD Jr. Materials Science and Engineering. 7th ed. NJ, USA: John Wiley & Sons; 2007

[2] Friedel J. Les Dislocations. Paris, France: g-V; 1956

[3] Zhuang Z, Liu Z, Cui Y. Dislocation Mechanism-Based Crystal Plasticity. Beigin, China: Elsevier; 2019

[4] Tomimura K, Takaki S, Tokunaga Y. Reversion mechanism from deformation induced martensite to austenite in metastable austenitic stainless steels. ISIJ International. 1991;**31**(12):1431-1437

[5] Zhilyaev AP, Oh-Ishi K, Langdon TG, McNelley TR. Microstructural evolution in commercial purity aluminum during high-pressure torsion. Materials Science and Engineering: A. 2005;**410**:277-280

[6] Niue LL, Zhang Y, Shu X, Gao F, Jin S, Zhou H-B, et al. Shear-coupled grain boundary migration assisted by unusual atomic shuffling. Scientific Reports. 2016;**6**:23062

[7] Torizuka S, Umezawa O, Tsuzaki K, Nagai K. Shape, size and crystallographic orientation of the ferrite grains formed at grain boundaries of deformed austenite in a low carbon steel. Tetsu to Hagane. 2000;**86**(12):807-814

[8] Rosochowski A. Severe Plastic Deformation Technology. Scotland, UK: Whittles Publishing; 2013

[9] Aizawa T, Tokumitsu K. Bulk mechanical alloying for productive process of functional alloys. Materials Science Forum. 1999;**312**:13-22

[10] Aizawa T. Functionalization of stainless steels via low temperature plasma nitriding. In: Proceedings of the 7th Annual Basic Science International Conference. Malang, Indonesia; 2017. pp. 1-16

[11] Farghali A, Aizawa T. Phase transformation induced by high nitrogen content solid solution in the martensitic stainless steels. Materials Transactions. 2017;**58**:697-700

[12] Farghali A, Aizawa T. Nitrogen supersaturation process in the AISI420 martensitic stainless steels by low temperature plasma nitriding. ISIJ International. 2018;**58**(3):401-407

[13] Aizawa T. Low temperature plasma nitriding of austenitic stainless steels. In: Stainless Steels. London, UK: IntechOpen; 2018. pp. 31-50

[14] Aizawa T, Yoshihara S. Homogeneous and heterogeneous micro-structuring of austenitic stainless steels by the low temperature plasma nitriding. In: Proceedings of 7th International Conference on Material Strength and Applied Mechanics. Kita-Kyushu, Japan; 2018

[15] Aizawa T. Characteristics of nitrided layer in dies – Low temperature plasma nitrided stainless steels. Bulletin of the JSTP. 2019;**2**:411-415

[16] Chen Z, Yang Y, Jiao H. Some Applications of Electron Back Scattering Diffraction (EBSD) in Materials Research. London, UK: IntechOpen; 2012. DOI: 10.5772/3526

[17] Aizawa T, Shiratori T, Komatsu T. Integrated manufacturing of fine-grained stainless steels for industries and medicals. In: High Entropy Alloy. London, UK: IntechOpen; 2019. pp. 121-143

[18] Takagi S, Tanimoto S, Tomimura K, Tokunaga Y. Optimal chemical composition in Fe-Cr-Ni

alloys for ultra grain refining by reversion of deformation induced martensite. Tetsu to Hagane. 1988;**74**(6):1052-1057

[19] Tomimura K, Takagi S, Tokunaga Y. Reversion process of deformation induced martensite to austenite in metastable austenitic stainless steels. Tetsu to Hagane. 1988;**74**(8):1649-1655

[20] Katoh M, Torisaka Y. Thermo-mechanical treatment for improvement of superplasticity of SUS304. Tetsu to Hagane. 1998;**84**(2):127-130

[21] Torizuka S, Muramatsu E, Murty SN, Ngai K. Microstructure evolution and strength-reduction in area balance of ultrafine-grained steels processed by warm caliber rolling. Scripta Materialia. 2006;**55**(8):751-754

[22] Shiratori T, Suzuki Y, Nakano S, Katoh M, Yang M, Komatsu T. Influence of grain size on sheared surface in micropiercing. The Japan Society for Technology of Plasticity. 2015;**56**(652):401-406

[23] Shiratori T, Suzuki Y, Nakano S, Yang M, Komatsu T. Effects of grain size on the sheared surface in micropiercing of austenitic stainless steel SUS304. Manufacturing Review. 2015;**2**:9

[24] Shiratori T, Katoh M, Satoh N, Yoshino T, Nakano S, Yang M. Deformation and transformation behavior in micropiercing of fine-grained SUS304. The Japan Society for Technology of Plasticity. 2017;**58**(681):936-942

[25] Tamura I. On the TRIP steel. Tetsu to Hagane. 1970;**56**(3):429-445

[26] Shrinivas V, Varma SK, Murr LE. Deformation-induced martensitic characteristics in 304 and 316 stainless steels during room temperature rolling. Metallurgical and Materials Transactions A. 1995;**26A**:661-671

[27] Ameyama K. Low temperature recrystallization and formation of an ultra-fine (γ+α) microduplex structure in a SUS316L stainless steel. Scripta Materialia. 1998;**38**(3):517-522

[28] Shiratori T. Effects of grain size on strain induced martensitic phase after shearing deformation in micro punching at SUS316L steel. In: Proceedings of the 174th ISIJ Meeting. Vol. 30. Sapporo, Japan; 2017. p. 963

[29] Aizawa T, Yoshino T, Shiratori T, Yoshihara S-I. Grain size effect on the nitrogen super-saturation process into AISI316 at 623 K. ISIJ International. 2019;**59**:1886-1892

[30] Aizawa T, Shiratori T, Yoshino T, Suzuki Y, Komatsu T. Microstructure evolution of fine-grained AISI316 wire plasma nitrided at 623 K through uniaxial tensile loading. ISIJ International. 2020. (in press)

[31] Thomason PF. Ductile Fracture of Metals. Oxford, UK: Pergamon Press; 1990. p. 92

[32] Takahashi M, Nagumo M, Asano I. Microstructures dominating the ductility of eutectoid pearlitic steels. Journal of the Japan Institute of Metals. 1978;**42**(7):708-715

[33] Cuevas AC, Bercerril EB, Martinez MS, Riuz JL. Metal matric composites. Switzerland: Springer Nature; 2018. p. 221

[34] Tsurui T, Inoue S, Matsuda K, Ishigaki H, Murata K, Koterasawa K. Effect of tensile axis on α' martensitic formation of SUS304 stainless steel single crystals. Journal of the Society of Materials Science. 2001;**50**(10):1115

[35] Zakhariev Z, editor. Polycrystalline Materials. London, UK: IntechOpen; 2012

[36] Sutton AP, Balluffi RW. Overview on geometric criteria for low interfacial energy. Acta Metallurgica. 1987;**35**(9):2177-2201

[37] Ouchi C, Okita T. Austenitic grain refinement through static recrystallization immediately after hot rolling. Transactions of the Iron and Steel Institute of Japan. 1984;**24**:726-733

[38] Kalpakjian S, Schmid SR. Manufacturing Engineering and Technology. 5th ed. NJ, USA: Pearson Prentice Hall; 2006

[39] Aizawa T, Yohino T. Plastic straining for microstructure refinement in stainless steels by low temperature plasma nitriding. In: Proc. 12th SEATUC Conference. YagYagarta, Indonesia; 2018

[40] Wang ZW, Palmer RE. Atomic-scale structure analysis by advanced transmission electron microscopy. Frontiers of Nanoscience. 2015;**9**:127-159

[41] Tarui T, Yamasaki S. Microstructure and mechanical properties of high carbon steel wire. Bulletin of the JSTP. 2018;**1**(2):111-115

[42] Kajino S. Microstructure and mechanical properties of surface layer in drawn wire. Bulletin of the JSTP. 2018;**1**(2):116-120

[43] Domain C, Becquart CS, Foct J. Ab initio study of foreign interstitial atom (C, N) interactions with intrinsic point defects in α-Fe. Physical Review B. 2004;**69**:144122

Chapter 6

Strongly Fluorescent Heterocyclic Molecule: Crystallography, 3D Hydrogen-Bonded, Fluorescence Study and QTAIM/TD-DFT/MESP Theoretical Analysis

Ouahida Zeghouan, Seifeddine Sellami and Mohamed AbdEsselem Dems

Abstract

In this chapter we explored the fluorescence properties of the title compound 1–10 phenanthroline hydrate (phh), $\{(C_{12}N_2H_8)\cdot H_2O\}$. The structure of phh is stabilized by strong as well as weak intermolecular interactions in the crystal. These interactions O—H···O, O—H···N, C—H···O and C—H···N hold the crystal structure in a three-dimensional network. Optical analysis (fluorescence) was performed on the test compound. The measurements in solvents of different polarities were carried out at ambient temperature (298 K). These results prompted us to investigate some photoluminescence applications for heterocyclic compounds as the sensing of blue-light luminescent materials. The time-dependent density functional theory (TD-DFT) calculations were performed on this compound, with the purpose to identify the origin of absorption and emission band, the nature of the electronic transitions. The atoms in molecules (AIM) theory and orbital analysis and molecular electrostatic potential (MESP) were applied to analyze the electron densities, their properties and the energy diagram of the molecular orbitals. The AIM and MESP analysis have been applied for part B of phh to demonstrate that the O1W—H11W···N1B type of interaction has the strongest hydrogen bond.

Keywords: aromatic molecule, X-ray diffraction, fluorescence, QTAIM/TD-DFT/MESP theoretical analysis

1. Introduction

The study of the photochemical and photophysical properties of heterocyclic compounds has received a great deal of attention during the last decade. 1,10-phenanthroline hydrate is a heterocyclic organic compound, used as a ligand in coordination chemistry; it has been the object of numerous studies, owing to its excellent complexing properties on metal ions. The multitude of applications of this cation motivated large development in synthesis of phenanthroline [1]. Various physico-chemical and biochemical techniques including UV/visible, fluorescence and viscometric titration, thermal denaturation and differential pulse voltammetry have been employed for 1,10-phenanthroline complexes to probe the details of DNA binding [2, 3]. The influence of the presence of nitrogen atoms on the fluorescence spectral maxima of aromatic molecules is our interest, including comparison of the spectra in polar and non-polar solvents. Such studies facilitate to understand the features of chemical bonding in molecules from the topological analysis of electron densities [4] and various electrostatic properties of molecules. These fundamental properties are directly related to the properties of materials [5]. The present compound, viz. $\{(C_{12}N_2H_8)\cdot H_2O\}$, has been characterized with X-ray crystallography, and we are interested in fluorescence properties. The fluorescence spectra of the present compound have been obtained in various conditions. The time-dependent density functional theory (TD-DFT)/quantum theory of atoms in molecules (QTAIM)/orbital analysis and molecular electrostatic potential (MESP) theoretical calculations have been applied too.

2. Structural commentary

The structure of $\{(C_{12}N_2H_8)\cdot H_2O\}$ is trigonal; space group P31, the molecular structure, is shown in **Figure 1**. The asymmetric unit is formed from three molecules of phenanthroline and three molecules of water. The crystal structure is built of successive rings formed by six molecules of phenanthroline; the centre of these

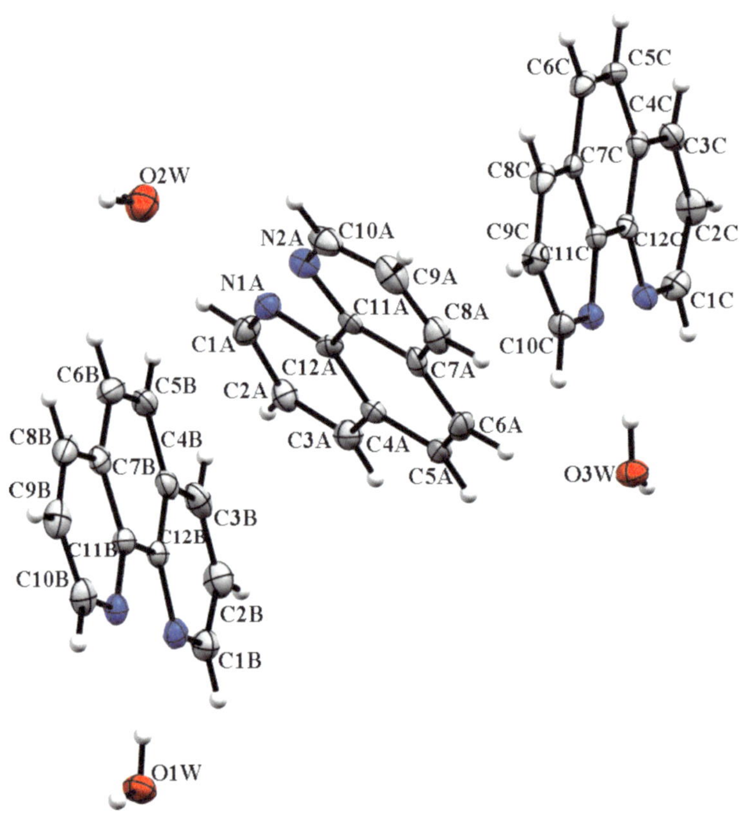

Figure 1.
View of the hpp molecule, with the atom-numbering scheme. Displacement ellipsoids are drawn at the 50% probability level, and H atoms are shown as circles of arbitrary radii.

rings are the three molecules of water. Bond lengths and valence angles compare well with the average values from related phenanthroline structures [6, 7].

3. Supramolecular features

The crystal structure is built of successive rings of phenanthroline and water molecules extending parallel to (001) plane that are connected by an extensive three-dimensional hydrogen-bonded network of the type O—H···O, O—H···N, C—H···O and C—H···N (**Table 1** and **Figure 2a**). Aqua molecules and N groups are involved in hydrogen bonding and form a three-dimensional network of infinite chains and variable degrees of rings [$C^3_3(7)$, $R^2_1(5)$, $R^2_2(9)$, $R^5_6(22)$ and $R^4_6(15)$] which deploy along the two crystallographic axes a and b (**Figure 2b** and **c**) [8].

4. Luminescent properties

Photoluminescence spectra were measured using a Cary Eclipse (Agilent Technologies) fluorescence spectrophotometer in a quartz cell (1.1 cm cross section) equipped with a xenon lamp and a dual monochromator. The measurements in solvents of different polarities were carried out at ambient temperature (298 K) with the slit ex/em = 20/10 nm. The photoluminescence properties of {$(C_{12}N_2H_8)\cdot H_2O$} in any solution were investigated in the visible region. The emission spectra of phh (10^{-4} M) in non-polar solvents, viz. benzene and toluene, show maximum at ~420 nm at λ_{ex} = 320 nm (**Figure 3**). Studies at higher concentrations of the probe were not possible as it is marginally soluble in non-polar solvents. In polar solvents, viz. methanol, ethanol and acetonitrile, the solution exhibit emission maximum at

D—H···A	D—H	H···A	D···A	D—H···A
O1W—H1W···O1Wi	0.8 (5)	2.2 (4)	2.95 (4)	163.00
O2W—H2W···N1A	0.9(3)	2.1 (2)	2.92 (4)	152.00
O2W—H2W···N2A	0.9(3)	2.5 (2)	3.18 (4)	135.00
O3W—H3W···O3Wii	0.84(19)	2.1 (2)	2.96 (4)	174.00
O1W—H11W···N1B	1.10(19)	1.9 (2)	2.90 (3)	152.00
O1W—H11W···N2B	1.10(19)	2.4 (3)	3.18 (3)	125.00
O2W—H22W···O2Wiii	0.8(2)	2.1 (2)	2.94 (3)	169.00
O3W—H33W···N1C	0.86(18)	2.5 (2)	3.16 (3)	138.00
O3W—H33W···N2C	0.86(18)	2.2 (2)	2.97 (3)	148.00
C1B—H1B···O1Wi	0.9300	2.5900	3.44 (5)	152.00
C3A—H3A···N1Civ	0.9300	2.4900	3.28 (5)	142.00
C8C—H037···N2Bv	0.9200	2.4700	3.31 (5)	151.00
C5B—H5B···N2A	0.9300	2.5700	3.41 (5)	151.00
C8B—H8B···N2Avi	0.9300	2.5000	3.25 (5)	139.00

Symmetry codes: (i) −y + 1, x − y + 1, z + 1/3; (ii) −x + y + 1, −x + 1, z − 1/3; (iii) −y, x − y, z + 1/3; (iv) −x + y + 1, −x + 1, z + 2/3; (v) −x + y, −x, z − 1/3; (vi) −y, x − y, z − 2/3.

Table 1.
Hydrogen bond geometry (Å, °).

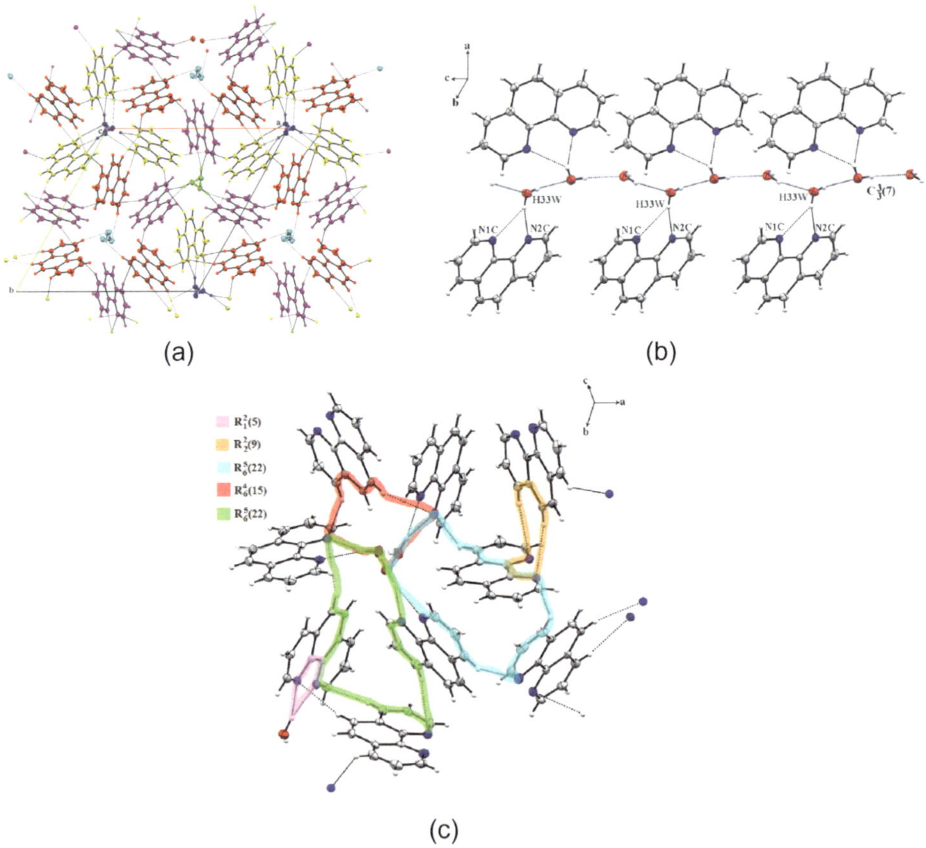

Figure 2.
(a) Partial view of the packing in the phh, showing the O—H O, O—H N, C—H O and C—H N hydrogen bonds. (b) Partial view of the packing in the phh, showing the formation of the infinite chain $C^3{}_3(7)$*. (c) Graph set* $R^2{}_1(5)$, $R^2{}_2(9)$, $R^5{}_6(22)$ *and* $R^4{}_6(15)$ *motifs. Atoms are represented as circles of different radii. Hydrogen bonds are shown as dashed lines.*

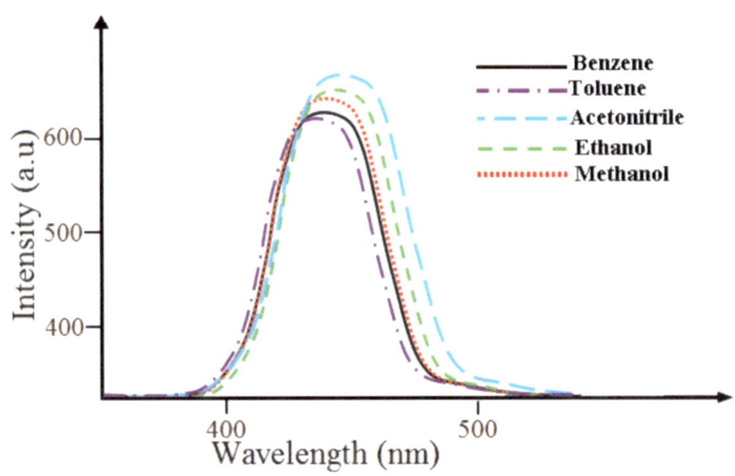

Figure 3.
Fluorescence spectra of phh in polar and non-polar solvent.

~440 nm, which resembles with the emission spectrum in non-polar solvents. This also indicates the absence of any aggregates or other conformers in the ground state. The studied compound was found to be fluorescently active in solution at room temperature showing an intense blue fluorescence with high fluorescent quantum yield, independent of the excitation wavelength in different solvents. The spectra clearly reveal that the emission originate from an excited state-charge transfer state π-π* because of the presence of N atoms. Thus, the compound phh may be a candidate for blue-light luminescent materials.

5. Theoretical analysis

All geometry optimizations of the molecular structure of phh were performed within the Amsterdam Density Functional (ADF) software [9], and all of theoretical calculations were carried out by the density functional theory (DFT), using the Perdew, Burke and Ernzerhof's exchange functional along with generalized gradient approximations, exchange and correlation functional GGA (PBE) [10, 11], employing the triple-zeta polarized (TZP) basis set.

Singlet excited states were optimized using time-dependent DFT (TD-DFT) calculations [12–15].

The geometric optimization has been carried on the molecular structure of phh. The ground state geometry was adapted from the X-ray data. Calculated structural parameters reveal a good agreement with the original X-ray diffraction data, which indicates significant stability of both compounds (**Table 2**).

5.1 TD-DFT absorption spectra

The time-dependent density functional theory calculations were performed on this compound, with the aim to identify the nature of the electronic transitions.

N1A—C1A	1.32 (4)	N2B—C10B	1.34 (5)
N1A—C12A	1.36 (4)	N2B—C11B	1.36 (4)
N2A—C10A	1.32 (4)	N1C—C12C	1.35 (4)
N2A—C11A	1.36 (4)	N1C—C1C	1.33 (4)
N1B—C12B	1.36 (4)	N2C—C11C	1.36 (4)
N1B—C1B	1.32 (5)	N2C—C10C	1.33 (4)
C1A—N1A—C12A	117 (3)	N1B—C1B—C2B	124 (3)
C10A—N2A—C11A	117 (3)	N2B—C10B—C9B	123 (3)
C1B—N1B—C12B	119 (3)	N2B—C11B—C7B	123 (3)
C10B—N2B—C11B	117 (3)	N2B—C11B—C12B	118 (2)
C1C—N1C—C12C	117 (2)	N1B—C12B—C11B	119 (3)
C10C—N2C—C11C	117 (3)	N1B—C12B—C4B	122 (3)
N1A—C1A—C2A	124 (3)	N1C—C1C—C2C	124 (3)
N2A—C10A—C9A	125 (4)	N2C—C10C—C9C	125 (3)
N2A—C11A—C7A	123 (3)	N2C—C11C—C7C	123 (3)
N2A—C11A—C12A	118 (3)	N2C—C11C—C12C	118 (2)
N1A—C12A—C4A	123 (3)	N1C—C12C—C4C	124 (3)
N1A—C12A—C11A	118 (3)	N1C—C12C—C11C	118 (2)

Table 2.
Selected geometric parameters (Å,°).

The absorption spectra of the studied compound are shown in **Figure 1**. The computed absorption bands, dominant transitions, characters and oscillator strengths (f) are given in **Table 3**.

The compound phh that absorbs in the region (λ = 165–280 nm) appears in **Figure 4**, and it shows three peaks with maxima at λ = 190, 225 and 268. **Table 3** clearly indicates that the lowest energy and the intense peak is at λ = 268 nm (E = 4.62 eV), transition character as corresponding it results from a transition H−2 to L+1. From **Figure 4** we observe two absorption peaks positioned at λ = 225 and 190 nm; the absorption bands correspond to the H−1 to L+3 and H−3 to L+2, respectively; and these transitions are affected to a π-π^* transition.

5.2 AIM topological analysis

Bader and Essen have shown that that the negative value of $\nabla^2(\rho)$ indicates that there is an electronic charge concentration at the BCP, which implies that the nature of the bond is covalent. For intermolecular interactions, is less weak and $\nabla > 0$, which implies the presence of a hydrogen-bonding interaction [16, 17].

The atoms in molecules (AIM) theory for part B (C1B to C12B/O1W) was applied to analyze the electron densities and their properties; details obtained from this approach are shown in **Table 4**. **Figure 5** shows the existence of a bond critical point between all of each two bonded atoms in part B.

We notice from **Table 4**, the values $\rho > 0.2$ and $\Delta < 0$ indicate that covalent bonds are present. And for the bond between N1B and H11W, we notice that $\rho = 0.034$ is very weak and $\Delta = 0.091 > 0$ (see **Table 4**) which indicate the presence of a strong hydrogen-bonding interaction. These results are in good agreement with those obtained by experimental refinement.

λ (nm)	f	E (ev)	Transition	Character
268	0.48	4.62	H−2 to L+1	π-π^*
225	0.47	5.49	H−1 to L+3	π-π^*
190	0.20	6.49	H−3 to L+2	π-π^*

Table 3.
The calculated optical transition energies and their corresponding oscillator strengths (f) for phh.

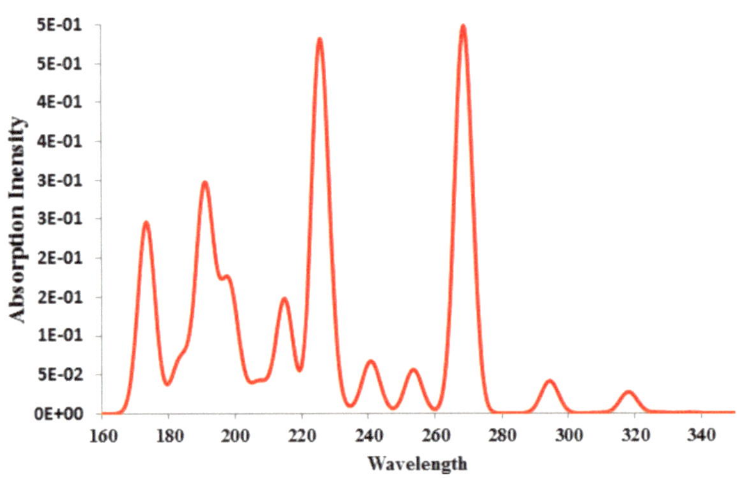

Figure 4.
TD-DFT absorption spectra of phh.

Basins	ρ	∇	|V|/G	H
N1B—H11W	0.034	0.091	1.063	−0.001
N1B—C1B	0.329	−0.869	3.275	−0.387
N1B—C12B	0.341	−0.946	3.138	−0.444
O1W—H1W	0.346	−1.686	6.984	−0.506
O1W—H11W	0.327	−1.685	7.426	−0.499

Table 4.
Results of atoms in molecule approach for part B of phh molecule.

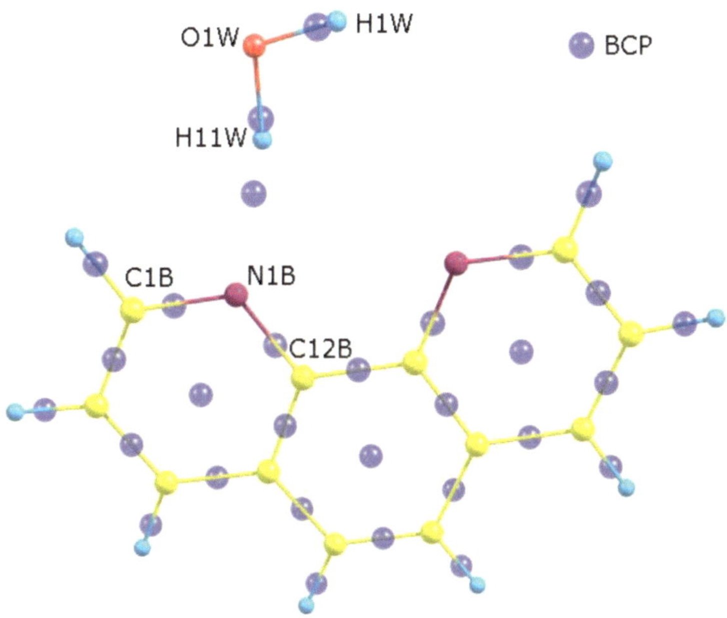

Figure 5.
The critical point founded between every pair of atoms in part B.

5.3 Orbital analysis and molecular electrostatic potential

The energy diagram of the molecular orbitals obtained by DFT approach for part B of compound phh is shown in **Figure 6**.

The HOMO and LUMO are important orbitals in a molecule; the gap between the occupied and vacant orbitals (HOMO/LUMO) is remarkable in the compound (2.92 eV; see **Figure 6**) and predicts its stability.

The highest occupied molecular orbital (HOMO) is mainly delocalized on the oxygen atom (93%), while it is found that the carbons of the cycles contribute strongly in the composition of the LUMO orbital with a participation rate of 88%.

The molecular electrostatic potential is a very important element for the illustration and visualization of the charge region variability and allows identifying the electrophilic and nucleophilic attack sites as well as the hydrogen-bonding interactions [18–21].

The blue region represents positive MESP, and the red region represents negative ESP, while the green region refers to the neutral region. From the MESP surface diagram, it can be seen that negative regions (red) are mainly localized over the oxygen and the nitrogen atom, while the positive regions (blue) are distributed over

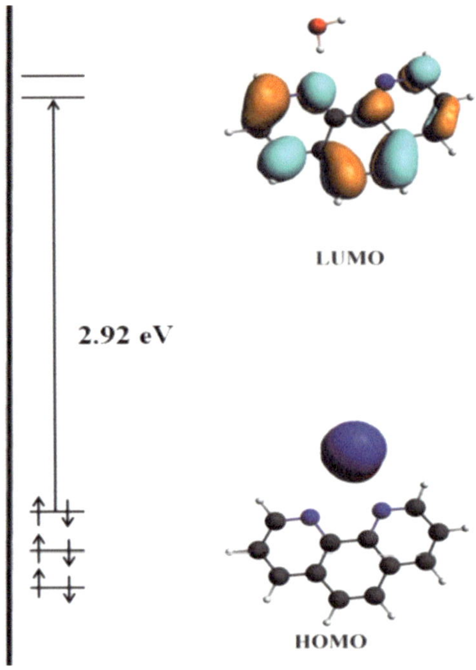

Figure 6.
Frontier molecular orbital diagram of part B.

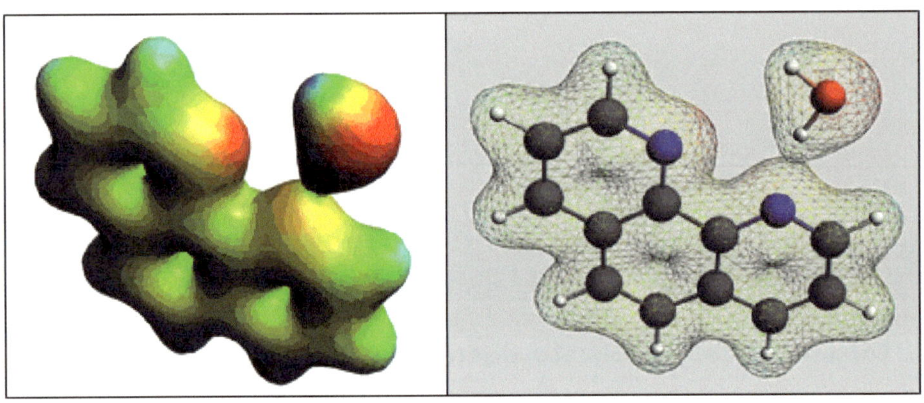

Figure 7.
MEP surface diagram of part B.

the H atoms. **Figure** 7 shows the presence of a hydrogen-bonding interaction between N1B and H11W.

6. Crystallization

On a Perkin-Elmer spectrometer, infrared spectra were recorded at room temperature in an interval of 500–4000 cm^{-1}.

The compound was crystallized by dissolving the 1–10 phenanthroline hydrate (1 g) in mixture solution of ethanol-water (V/V = 1:1). The solution was preserved at room temperature under agitation during 4 h. After a slow evaporation in the interior of 2 days, transparent prism crystals were obtained.

The crystals formed were washed and filtered using 20 ml of water.

The most important infrared wavelength (cm^{-1}):

(1): 3400 (vs), 3084 (m), 1632 (w), 1617 (w), 1575 (s), 1564 (m), 1502 (s), 1443 (w), 1414 (vs), 1340 (w), 1285 (vw), 1208 (w), 1132 (m), 1082 (w), 1031 (w), 987 (m), 860 (vw), 764 (w), 733 (vs).

7. Programmes

Computer programmes by Bruker *APEX2* (2006) [22]; *SAINT* (Bruker, 2006) [22]; *SAINT*, *SIR2002* [23], and *SHELXL2008* [24]; *WinGX* [25]; *Mercury* Version 1.4 [26]

8. Refinement

Crystal data, data collection and structure refinement details are summarized in **Table 5**. H atoms were placed at calculated positions with C—H = 0.93 Å (aromatic

	(hpp)
	Crystal data
Chemical formula	C13.50H11.25N2.25O1.13
Mr	223.00
Crystal system, space group	Trigonal, P31
Temperature (K)	293
a, c (Å)	17.5075 (7), 8.4300 (4)
V (Å^3)	2237.72 (14)
Z	8
Radiation type	Mo Kα
μ (mm^{-1})	0.09
Crystal size (mm)	0.2 × 0.1 × 0.08
	Data collection
Diffractometer	Bruker APEX-II CCD
Absorption correction	—
No. of measured, independent and observed [I > 2σ(I)] reflections	13,243, 5115, 4195
Rint	0.046
(sin θ/λ)max (Å^{-1})	0.596
	Refinement
R[F2 > 2σ(F2)], wR(F2), S	0.047, 0.098, 1.06
No. of reflections	5115
No. of parameters	418
No. of restraints	10
Δρmax, Δρmin (e Å^{-3})	0.25, −0.19
Absolute structure	[27]
Absolute structure parameter	0.4 (15)

Table 5.
Experimental details.

H atoms) and refined in riding mode with $U_{iso}(H) = 1.2U_{eq}(C)$. The O-bound H atoms were located in a Fourier map and refined with O—H restraint of 0.85 (1)Å ($U_{iso}(H) = 1.5U_{eq}(O)$).

9. Conclusion

The experimental and theoretical charge-density analysis has been performed to understand the topological and electrostatic properties of the 1–10 phenanthroline hydrate molecule. The crystal structure reveals the molecule forming O—H···O, O—H···N, C—H···O and C—H···N types of interactions with the neighboring molecules. Furthermore, photoluminescence studies indicate that the hpp is fluorescently active in solution at room temperature and reveal that the emission originate from an excited state-charge transfer state π-π^* which is proven by TD-DFT analysis.

The AIM analysis shows the existence of a bond critical point (BCP) between N1B and H11W. We notice that $\rho = 0.034$ is very weak and $\Delta = 0.091 > 0$ which confirms that this interaction is the strongest interaction in the molecule (hydrogen-bonding interaction). These results are in good agreement with those obtained by experimental refinement.

The structural, topological and electrostatic properties of the 1–10 phenanthroline hydrate molecule obtained from the X-ray diffraction method may be useful to design a new candidate of blue-light luminescent materials.

Acknowledgements

This work was supported by the Biotechnology Research Center (CRBt), Constantine, Algérie.

Author details

Ouahida Zeghouan[1*], Seifeddine Sellami[2] and Mohamed AbdEsselem Dems[1]

1 Biotechnology Research Center (CRBt), Constantine, Algeria

2 Laboratoire Pollution et Traitement des Eaux, Département de Chimie, Faculté des Sciences Exactes, Université Frères Mentouri Constantine 1, Constantine, Algeria

*Address all correspondence to: ouahida.zeghouan@gmail.com

References

[1] Sastri CV, Eswaramoorthy D, Giribabu L, Maiya BG. DNA interactions of new mixed-ligand complexes of cobalt(III) and nickel(II) that incorporate modified phenanthroline ligands. Journal of Inorganic Biochemistry. 2003;**94**: 138-145

[2] Cohen G, Eisenberg H. Viscosity and sedimentation study of sonicated DNA–proflavine complexes. Biopolymers. 1969;**8**:45-55

[3] Satyanarayana S, Dabrowiak JC, Chaires JB. Neither DELTA- nor LAMBDA-tris(phenanthroline) ruthenium (II) binds to DNA by classical intercalation. Biochemistry. 1992;**31**:9319-9324

[4] Bader RFW. Atoms in Molecules: A Quantum Theory. Oxford, New York: Clarendon Press; 1995

[5] Coppens P. X-Ray Charge Density and Chemical Bonding. Oxford University Press; 1997

[6] Gropper H, Doerrr F. Berichte der Bunsengesellschaft für physikalische Chemie. 1963;**67**:46-54

[7] Shulman SG. The IR spectrum of 4,7-phenanthroline in the region 650-900 cm-1 has been analyzed, and the C—H ... 77. Fluorescence News. 1973;7:33

[8] Bernstein J, Davis RE, Shimoni L, Chang N-L. Patterns in Hydrogen Bonding Functionality and Graph Set Analysis in Crystals. Angewandte Chemie (International Ed. in English). 1995;**34**:1555-1573

[9] Baerends EJ, Ellis DE, Ros P. Self-consistent molecular Hartree—Fock—Slater calculations I. The computational procedure. The Journal of Chemical Physics. 1973;**2**(1):41-51

[10] Perdew JP, Burke K, Ernzerhof M. Generalized gradient approximation made simple. Physical Review Letters. 1997;**78**:1396-1396

[11] Langreth DC, Vosko SH. In density functional theory of many-fermion systems. Advances in Quantum Chemistry. 1990;**21**:175-175

[12] Bauernschmitt R, Ahlrichs R. Stability analysis for solutions of the closed shell Kohn–Sham equation. The Journal of Chemical Physics. 1996;**104**: 9047-9052

[13] Gross EUK, Dobson JF, Petersilka M. Density functional theory of time dependent phenomena. In: Hafner K, Houk KN, Lehn IJM, Raymond KN, Rees CW, Thiem J, Vogtle F, editors. Topics in Current Chemistry—Density Functional Theory II. Berlin: Springer; 1996. pp. 81-172

[14] Casida ME. In: Chong PD, editor. Recent Advances in Density Functional Methods. Vol. 1. Canada: University of British Columbia; 1995. pp. 155-193. ISBN: 978-981-283-058-6

[15] Gross EKU, Kohn W. Time-dependent density-functional theory. Advances in Quantum Chemistry. 1990; **21**:255-291

[16] Bader RFW, Essén H. The characterization of atomic interactions. Journal of Chemical Physics. 1984;**80**: 1943

[17] Bader RFW, Matta CF. Bonding to titanium. Inorganic Chemistry. 2001;**40**: 5603

[18] Politzer P, Daiker KC. Models for chemical reactivity. In: Deb BM, editor. The Force Concept in Chemistry. New York: Van Nostrand Reinhold; 1981. pp. 294-387

[19] Politzer P, Laurence PR, Jayasuriya K, McKinney J. Molecular

electrostatic potentials: An effective tool for the elucidation of biochemical phenomena. In: McKinney J, editor. Structure Activity Correlation in Mechanism Studies and Predictive Toxicology, Environ. 1985;**61**:191-202

[20] Politzer P, Murray JS. Theoretical Biochemistry and Molecular Biophysics: A Comprehensive Survey. Vol. 2. Schenectady, NY: Adenine Press; 1991. pp. 165-191

[21] Scrocco E, Tomasi J. Topics in Current Chemistry. Vol. 42. Berlin: Springer-Verlag; 1973. p. 95

[22] Bruker. APEX2 and SAINT. Madison,Wisconsin, USA: Bruker AXS Inc.; 2006

[23] Burla MC, Camalli M, Carrozzini B, Cascarano GL, Giacovazzo C, Polidori G, et al. SIR prorram. Journal of Applied Crystallography. 2003;**36**:1103

[24] Sheldrick GM. Crystal structure refinement with SHELXL. Acta Crystallographica. 2008;**A64**:112-122

[25] Farrugia LJ. WinGX and ORTEP for Windows: An update. Journal of Applied Crystallography. 2012;**45**:849-854

[26] Macrae CF, Bruno IJ, Chisholm JA, Edgington PR, McCabe P, Pidcock E, et al. New features for the visualization and investigation of crystal structures. Journal of Applied Crystallography. 2008;**41**:466-470

[27] Flack HD. On enantiomorph-polarity estimation. Acta Crystallographica. 1983;**A39**:876-881

Chapter 7

Study on Dislocation-Dopant Ions Interaction during Plastic Deformation by Combination Method of Strain-Rate Cycling Tests and Application of Ultrasonic Oscillations

Yohichi Kohzuki

Abstract

Strain-rate cycling tests associated with the ultrasonic oscillation were conducted for the purpose of investigation on the interaction between dislocation and dopant ions during plastic deformation of seven kinds of single crystals: NaCl doped with Li^+, K^+, Rb^+, Cs^+, F^-, Br^- or I^- ions separately. Relative curves between the stress decrement ($\Delta\tau$) due to ultrasonic oscillatory stress and strain-rate sensitivity (λ) of flow stress under superposition of the oscillation are obtained by the original method (combination method of strain-rate cycling tests and application of ultrasonic oscillations) at 77 K to room temperature and have stair-like shapes for the specimens at low temperatures. The Gibbs free energy for overcoming of the dopant ion by dislocation at absolute zero is calculated from the data analyzed in terms of $\Delta\tau$ vs. λ. As a result, the obtained energies are found to be varied linearly with the isotropic defect around it in the each specimen.

Keywords: dislocation, ultrasonic oscillatory stress, activation energy, monovalent ion, isotropic strain

1. Introduction

Dislocation (linear defects in crystal) motions are related to the plasticity of crystal in a microscopic viewpoint. It is well known that the solution hardening depends on dislocation motion hindered by the atomic defects around impurities in crystals and is namely influenced by the dislocation-point defects interaction, which has been widely investigated by various methods. For instance, measurements of yield stress (e.g., [1–7]) and proof stress (e.g., [8, 9]), micro-hardness tests (e.g., [10–14]), direct observations of dislocation (e.g., [15–21]), internal friction measurements (e.g., [22–27]), or stress relaxation tests (e.g., [28, 29]) have been carried out so far. Nevertheless, it is difficult to obtain such information on the motion of the dislocation which moves by overcoming the forest dislocations and the weak obstacles such as impurities during plastic deformation of bulk. A large

IntechOpen

number of investigations have been conducted by the separation of the flow stress into effective and internal stresses on the basis of the temperature dependence of yield stress, the strain rate dependence of flow stress, and the stress relaxation. Yield stress depends on dislocation velocity, dislocation density, and multiplication of dislocations [30]. On the other hand, the effect of heat treatment on the microhardness is almost insensitive to the change of atomic order of point defects in a specimen. As for direct observations, electron microscopy provides the information on dislocation motion for a thin specimen but not for bulk, and also light scattering method is useful only for a transparent specimen. X-ray topography is the lack of resolution in the photograph, so that the specimen is limited to the low dislocation density below 10^4 cm^{-2}. Internal friction measurements concern the motion of the dislocation which breaks away from the weak obstacles between two forest dislocations by vibration [31]. Stress relaxation tests are generally assumed that internal structure of crystals does not change, i.e., dislocation density and internal stress are constant. Above-mentioned methods cannot provide the information on dislocation-obstacles interaction in bulk during plastic deformation.

In this chapter, the study on interaction between a dislocation and dopant ions is made by the strain-rate cycling tests during the Blaha effect measurement. The original method (strain-rate cycling tests associated with the Blaha effect measurement) is different from above-mentioned ones and would be possible to clear up it. The Blaha effect is the phenomenon that static flow stress decreases when an ultrasonic oscillatory stress is superimposed during plastic deformation [32]. Ohgaku and Takeuchi [33, 34] reported that the strain-rate cycling under the application of oscillation can separate the contributions arising from the interaction between a dislocation and dopant ions and from the dislocations themselves during plastic deformation at room temperature. Using ionic single crystals of KCl doped with Br^- (0.5, 1.0, and 2.0 mol%) or I^- (0.2, 0.5, and 1.0 mol%) [35] and of NaCl doped with Br^- (0.1, 0.5, and 1.0 mol%) [36], they discussed temperature dependence of the effective stress due to monovalent dopants (i.e., Br^- or I^-) and found that the measurement of strain-rate sensitivity under the ultrasonic oscillatory stress provides useful information on a mobile dislocation-the dopant ions interaction [35, 36]. The information on the dislocation motion breaking-away from dopant ions [37–40] and also X-irradiation induced defects [41] with the ultrasonic oscillatory stress has been successively provided by the original method, which seemed to separate the contributions arising from the dislocation-the point defects interaction and from dislocations themselves during plastic deformation of crystals.

The Blaha effect was found by Blaha and Langenecker when the ultrasonic oscillatory stress of 800 kHz was superimposed during plastic deformation of Zn single crystals. The same phenomenon as Zn crystals has been also observed in many metals (e.g., [42–44]). Since this phenomenon has a significance as an industrial purpose, it has been widely made to apply to the plastic working technique: wire drawing, deep drawing, rolling, and another metal forming techniques (e.g., [45–53]).

The strain-rate cycling tests associated with ultrasonic oscillation were carried out here for NaCl single crystals doped with various monovalent ions separately. The monovalent ion is considered to have isotropic strain in the alkali halide crystal because its size is different from the substituted ion of the host crystal. Dopant ions are expected to cause the hardening due to the dislocation motion hindered by the defects around them at low temperature. Its force-distance profile between a dislocation and an atomic defect is expressed by Cottrell and Bilby [54]. This chapter refers to the energy supplied by the thermal fluctuations, when the dopant ions are overcome by a dislocation with the help of thermal activation during plastic deformation of crystals. This is estimated from the dependence of the effective stress due to impurities on activation volume, which reveals the force-distance profile, given by the measurement of the stress decrement due to application of ultrasonic oscillatory stress and strain-rate sensitivity of flow stress under superimposition of ultrasonic oscillation. And further, it is presented that the difference in size of isotropic strain around the various dopants different from host ion has an influential factor of the energy for overcoming the dopant ion by a dislocation in several kinds of alkali-halide single crystals.

2. Combination method of strain-rate cycling tests and the Blaha effect measurement

Specimens used in this work were seven kinds of single crystals: NaCl doped with Li^+, K^+, Rb^+, Cs^+, F^-, Br^- or I^- ions separately. Each concentration of the dopants was 0.5 mol% in the melt. The specimens were prepared by cleaving the single crystalline ingots, which were grown by the Kyropoulos method [55] in air, to the size of 5 × 5 × 15 mm^3. Furthermore, they were kept immediately below the melting point for 24 h and were gradually cooled to room temperature at a rate of 40 K h^{-1}. This heat treatment was carried out for the purpose of reducing dislocation density as much as possible.

The schematic illustration of apparatus is shown in **Figure 1**. A resonator composed of a vibrator and a horn was attached to the testing machine, INSTRON Type 4465. The specimens were lightly fixed on a piezoelectric transducer and then cooled down to a test temperature. Each specimen was held at the test temperature for 30 min prior to the following test. The specimens were deformed by compression along the <100> axis at 77 K up to the room temperature, and the ultrasonic oscillatory stress was intermittently superimposed for 1 or 2 min by the resonator in the same direction as the compression. The temperature measurements of specimens were conducted by heater controlled using thermocouples of Ni-55%Cu vs. Cu. As for the tests at 77 K, the specimen was immersed in the liquid nitrogen. The stability of temperature during the test was kept within 2 K. The resonant frequency was 20 kHz from a multifunction synthesizer and the amplitude of the oscillatory

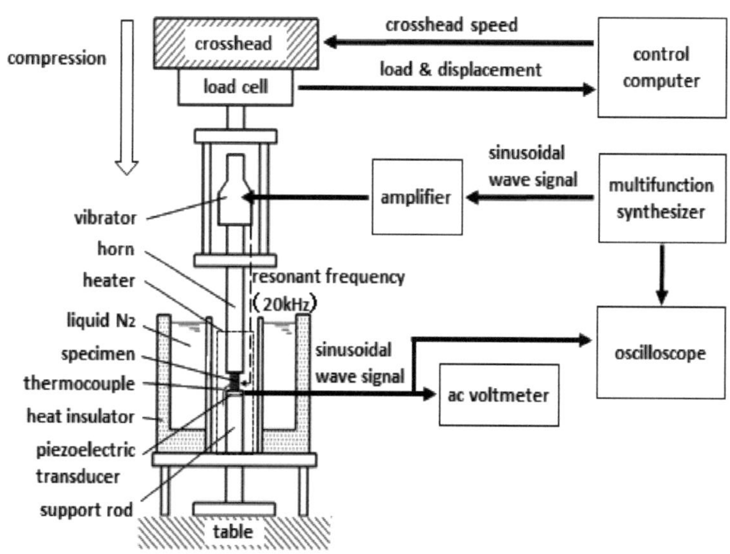

Figure 1.
Schematic block diagram of apparatus system.

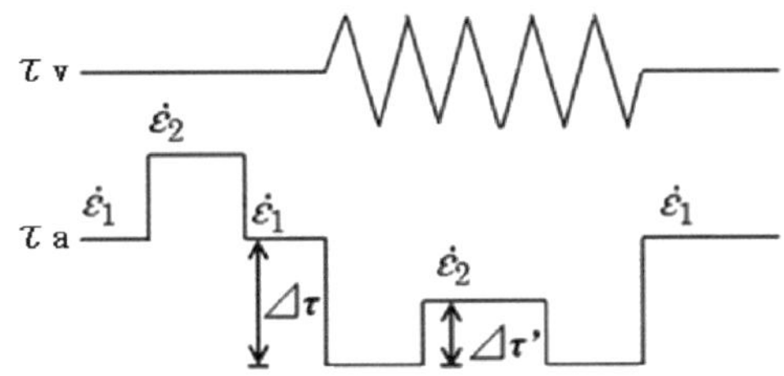

Figure 2.
Explanatory diagram of a change in applied shear stress, τ_a, for the strain-rate cycling test between the strain rates, $\dot{\varepsilon}_1$ ($2.2 \times 10^{-5}\ s^{-1}$) and $\dot{\varepsilon}_2$ ($1.1 \times 10^{-4}\ s^{-1}$), under superposition of ultrasonic oscillatory shear stress, τ_v.

stress was monitored by the output voltage from the piezoelectric transducer set between a specimen and the support rod, which was observed by an a.c. voltmeter or an oscilloscope. Since the wavelength, which is 226 mm on the basis of calculating from the data of ref. [56], is 15 times as long as the length of specimen, the strain of specimen is supposed to be homogeneous.

Strain-rate cycling tests made between the crosshead speeds of 10 and 50 μm $\min^{-1}$ were performed within the temperatures. The strain-rate cycling test associated with the ultrasonic oscillation is illustrated in **Figure 2**. Superposition of oscillatory stress (τ_v) causes a stress drop ($\Delta\tau$) during plastic deformation. When the strain-rate cycling between strain-rates of $\dot{\varepsilon}_1$ ($2.2 \times 10^{-5}\ s^{-1}$) and $\dot{\varepsilon}_2$ ($1.1 \times 10^{-4}\ s^{-1}$) was carried out keeping the stress amplitude of τ_v constant, the variation of stress due to the strain-rate cycling is $\Delta\tau$'. The strain-rate sensitivity ($\Delta\tau'/\Delta\ln\dot{\varepsilon}$) of the flow stress, which is given by $\Delta\tau'/1.609$, was used as a measurement of the strain-rate sensitivity ($\lambda = \Delta\tau'/\Delta\ln\dot{\varepsilon}$). Slip system for rock-salt structure such as NaCl crystal is {110} <1$\bar{1}$0> so that shear stress (τ) and shear strain (ε) calculated for the slip system were used in this study.

3. Relation between stress decrement ($\Delta\tau$) and strain-rate sensitivity (λ)

Figure 3 shows the influence of temperature on $\Delta\tau$ vs. λ curve for the NaCl:Rb^+ (0.5 mol%) single crystals at strain 6%. The variation of λ with $\Delta\tau$ has stair-like shape: two bending points and two plateau regions are on the each curve. That is to say, the first plateau region ranges below the first bending point at low $\Delta\tau$ and the second one extends from the second bending point at high $\Delta\tau$. λ gradually decreases with increasing $\Delta\tau$ between the two bending points. The length of $\Delta\tau$ within the first plateau region is named τ_p as denoted in **Figure 3**. τ_p tends to be lower at higher temperature. Similar phenomena as **Figure 3** are observed for all the other NaCl single crystals contained with the monovalent impurities (i.e. Li^+, K^+, Cs^+, F^-, Br^- or I^- ions).

The relation between $\Delta\tau$ and λ reflects the effect of ultrasonic oscillation on the dislocation motion on the slip plane containing many weak obstacles such as impurities and a few forest dislocations during plastic deformation [40]. $\Delta\tau$ vs. λ curve is divided into three regions as shown in **Figure 3**. Within the first plateau region of relative curve (i.e. region 1 in **Figure 3**), the application of oscillation with low stress amplitude cannot influence the average length of dislocation segments ($\bar{l}$) and $\bar{l}$ is considered to remain constant. All weak obstacles act as impedimenta to the dislocation motion there. In region 2, the dislocation begins to break-away from the weak ones between the forest dislocations by applying oscillation with high stress amplitude. As a result, $\bar{l}$ begins to increase and the λ of flow stress starts to decrease at the stress decrement $\Delta\tau$ of τ_p. This is because λ is inversely proportional to $\bar{l}$ [57].

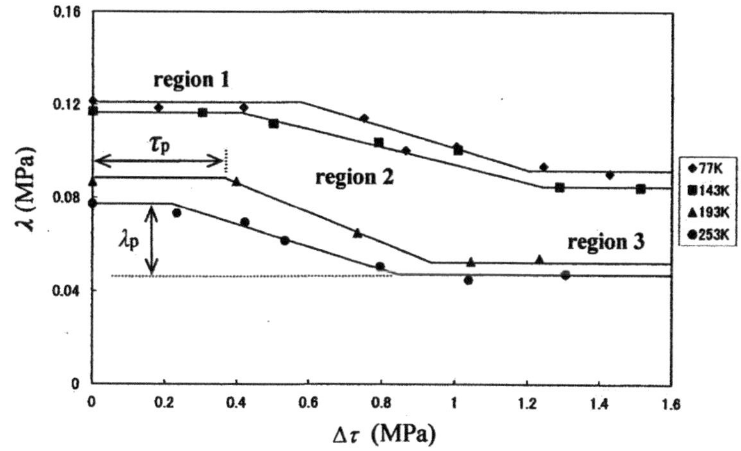

Figure 3.
Relation between the stress decrement (Δτ) and the strain-rate sensitivity (λ) for NaCl:Rb⁺ (0.5 mol%) at strain 6% and various temperatures. The numbers besides each symbol represent the temperature (reproduced from Ref. [58] with permission from the publisher).

Some weak obstacles stop acting as impedimenta in the region. The weak obstacles are supposed to be monovalent dopants (Li^+, K^+, Rb^+, Cs^+, F^-, Br^- or I^- ions) and not to be vacancies here, since the vacancies have low density as against the dopants in the specimen. When the specimens were plastically deformed, it is imagined that a dislocation begins to overcome from the dopants which lie on the dislocation with the help of thermal activation. Then, τ_p is considered to represent the effective stress due to the ions. Accordingly, τ_p is expected to decrease with increasing temperature. $\Delta\tau$ vs. λ curves shown in **Figure 3** correspond to this. As the temperature becomes larger, τ_p shifts in the direction of lower $\Delta\tau$. τ_p depends on type and density of the weak obstacle [36, 38]. Applying still larger stress amplitude during plastic deformation of the specimens, the second plateau region within stage 3 becomes to appear on the relative curves in **Figure 3**. In stage 3, the dopants are no longer act as the impedimenta to mobile dislocations and the dislocations are hindered only by forest dislocations. Then, $\bar{l}$ becomes constant again. This leads to the constant λ of flow stress. λ_p denoted in **Figure 3** is introduced later.

4. Model overcoming the thermal obstacle by a dislocation

A dislocation will encounter a stress field illustrated schematically in **Figure 4** as it moves through on the slip plane containing many weak obstacles and a few strong ones. In the figure, the positive stress concerning axis of the ordinate opposes the flow stress (applied stress τ) and the negative stress assists it. Extrinsic resistance to the dislocation motion has two types: one is long-range obstacle (the order of 10 atomic diameters or greater) and the other short-range obstacle (less than about 10 atomic diameters). The former is considered to be forest dislocations, large precipitates or second-phase particles, and grain boundary, for instance, and the latter impurity atoms, isolated and clustered point defects, small precipitates, intersecting dislocations, etc. Overcoming the latter type of obstacles (byname, thermal obstacles) by a dislocation, thermal fluctuations play an important role in aid of the flow stress above the temperature of 0 K. Then the aid energy, ΔG, supplied by the thermal fluctuations is given by the shaded part in **Figure 4**. Thus the dislocation can move through below τ_0 (i.e. effective stress τ^* due to short-range obstacles and internal stress τ_i

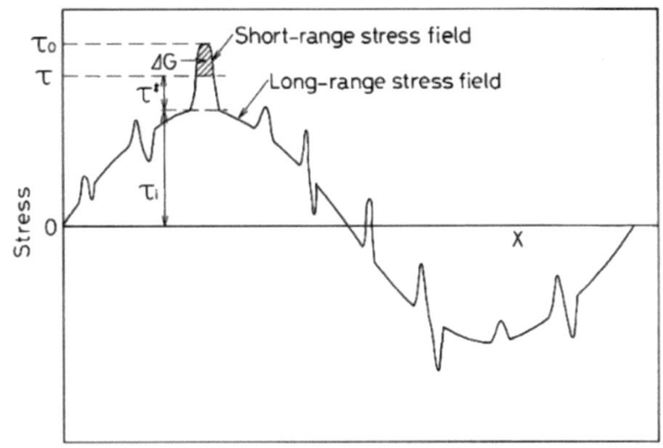

Figure 4.
Stress fields encountered by a dislocation moving through the crystal lattice [57].

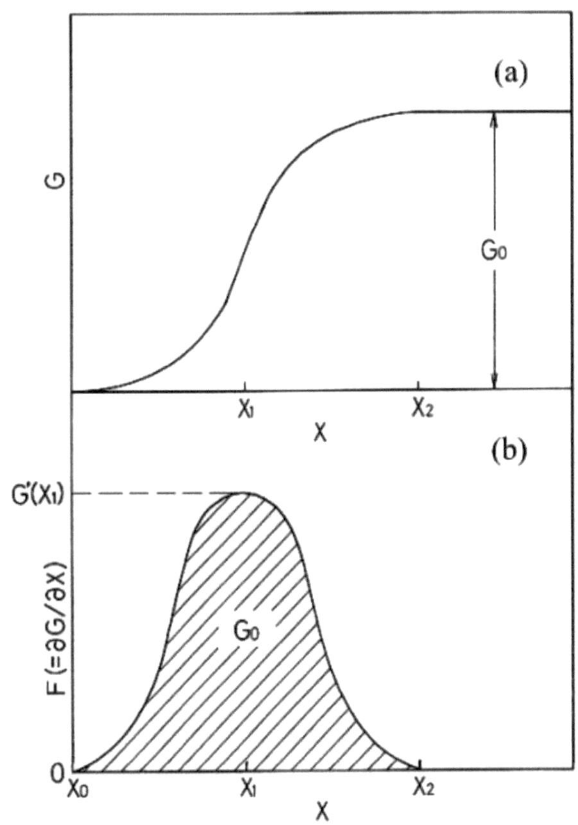

Figure 5.
The process for thermal activated overcoming of the short-range obstacle by a dislocation. Variation in (a) the Gibbs free energy of activation and (b) the force acted on the dislocation with the distance for a dislocation motion [57].

due to long-range ones in **Figure 4**). τ_0 is the value of τ at 0 K. As for the long-range obstacles (byname, athermal obstacles), the energy barrier is so large that the thermal fluctuations play no role in overcoming them within the temperature range.

The representation of **Figure 5** is concerned with a common type of thermal activation barrier. The free energy (G) varies with the distance (x) between a dislocation and the obstacle as given in **Figure 5(a)**. When a dislocation overcomes the short-range obstacles, the free energy becomes high on account of the work (ΔW) done by the applied stress. Then the resistance (F), where it can be defined by the differentiation of free energy with respect to x (i.e. $\partial G/\partial x$), to the dislocation motion is revealed as **Figure 5(b)** in accord with the abscissa of **Figure 5(a)**. F value is maximum at position x_1. **Figure 5(b)** corresponds to typical force-distance curve for short-range obstacle among those in **Figure 4**. Shape of this curve represented by $F(x)$ means the model overcoming the obstacle by a dislocation. G_0, which is taken as the shaded area under $F(x)$ between saddle-point positions x_0 and x_2 in **Figure 5(b)**, is the Gibbs free energy of activation for the breakaway of the dislocation from the obstacle in the absence of an applied stress (in this case it is equivalent to the Helmholtz free energy for the dislocation motion).

5. Relation between the effective stress due to impurities on activation volume

When the dislocation breaks-away from the defects on a slip plane with the aid of thermal activation during plastic deformation, observations of τ_p and λ_p would provide information on the dislocation-defect interaction in the specimen. λ_p is the difference between λ at first plateau place and at second one on $\Delta\tau$ vs. λ curve as presented in **Figure 3**, which has been regarded as the component of strain-rate sensitivity due to dopant ions when a dislocation moves forward with the help of oscillation [40].

Figure 6 shows the relation between τ_p and activation volume (V) for NaCl:Rb^+ (0.5 mol%). The activation volume has been expressed as [57].

$$V = kT\left(\frac{\partial \ln \dot{\varepsilon}}{\partial \tau}\right) \tag{1}$$

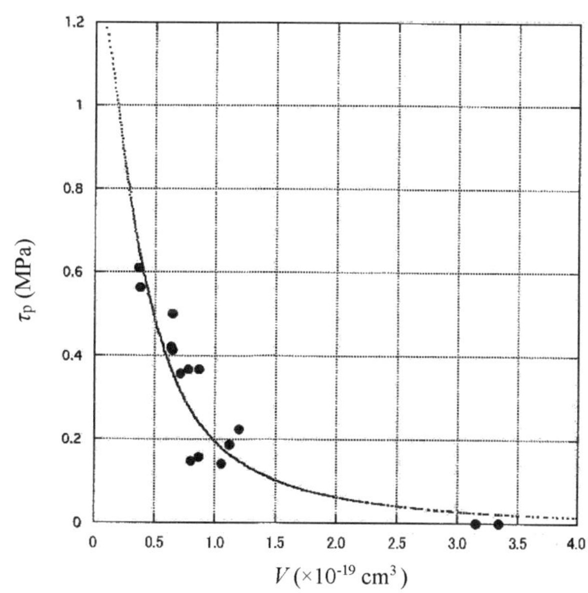

Figure 6.
Relation between τ_p and activation volume (V) for NaCl:Rb^+ (0.5 mol%) (reproduced from Ref. [58] with permission from the publisher).

Specimen	G_0 (eV)
NaCl:Li$^+$ (0.5 mol%)	0.55
NaCl:K$^+$ (0.5 mol%)	0.60
NaCl:Rb$^+$ (0.5 mol%)	0.61
NaCl:Cs$^+$ (0.5 mol%)	0.82
NaCl:F$^-$ (0.5 mol%)	0.69
NaCl:Br$^-$ (0.5 mol%)	0.47
NaCl:I$^-$ (0.5 mol%)	0.53

Table 1.
Values of energy G_o.

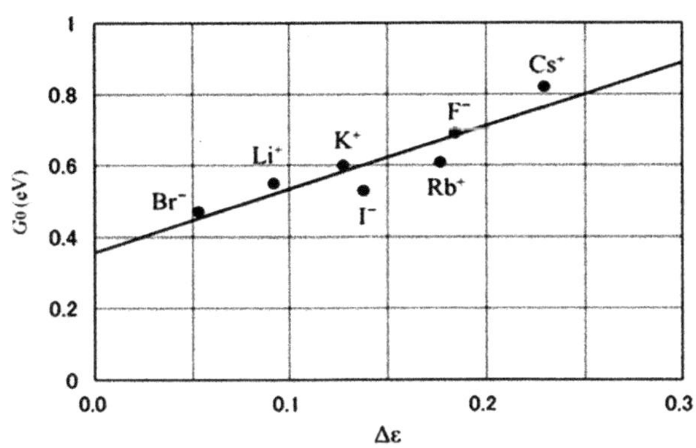

Figure 7.
Variation of the interaction energy (G_o) between dislocation and the dopant ion with the defect size (reproduced from Ref. [58] with permission from the publisher).

where k is the Boltzmann constant and T is the absolute temperature.

Here, the $\left(\frac{\partial \ln \dot{\varepsilon}}{\partial \tau}\right)$ in Eq. (1) is obtained from λ_p. Eq. (1) is namely replaced by

$$V = kT/\lambda_p. \quad (2)$$

This dependence (τ_p vs. V) also represents the force-distance profile between dislocation and Rb$^+$ ion. The τ_p vs. V curve gives the value of G_0 for the specimen. The G_0 values for the other specimens (i.e. NaCl: Li$^+$, K$^+$, Cs$^+$, F$^-$, Br$^-$ or I$^-$) are similarly estimated and are listed in **Table 1**.

Figure 7 shows the obtained energies G_0 with the isotropic defect size ($\Delta\varepsilon$), which is estimated from the difference between the lattice constants of host crystal and dopant, around ion doped in the each specimen. The ions beside each plot represent the dopants in NaCl single crystals. G_0 values vary linearly with $\Delta\varepsilon$ in the specimens. The intercept of the straight line is 0.36 eV, which is considered to be the interaction energy between dislocation and inherent obstacle of the host crystal because $\Delta\varepsilon$ is zero.

6. Conclusions

The following conclusions were derived from the data analyzed in terms of the $\Delta\tau$ vs. λ curves for NaCl: Li$^+$, K$^+$, Rb$^+$, Cs$^+$, Br$^-$, I$^-$, F$^-$, Br$^-$ or I$^-$ single crystals.

1. The relation between $\Delta\tau$ and λ has stair-like shape for the specimens at a given temperature and strain. There are two bending points and two plateau regions. λ decreases with $\Delta\tau$ between the two bending points. The measurement of τ_p and V calculated with λ_p provides information on the interaction between mobile dislocation and the dopant ion in the specimens during plastic deformation.

2. The Gibbs free energy G_0 for the overcoming of dislocation from the dopant is obtained for each of the specimens and increases linearly with increasing the defect size $\Delta\varepsilon$. This result leads to the conclusion that the dopant ion as weak obstacle to dislocation motion becomes slightly stronger with larger defect size around the dopant in NaCl single crystal.

Acknowledgements

Dr. T. Ohgaku, as well as S. Yamaguchi, M. Azuma, H. Teraji, E. Ogawa and Y. Yamanaka are acknowledged for his collaboration in the analysis on $\Delta\tau$ and λ data, as well as for their experimental assistance.

Conflict of interest

The author declares no conflict of interest.

Author details

Yohichi Kohzuki
Department of Mechanical Engineering, Saitama Institute of Technology, Fukaya, Japan

*Address all correspondence to: kohzuki@sit.ac.jp

References

[1] Chin GY, Van Uitert LG, Green ML, Zydzik GJ, Kometani TY. Strengthening of alkali halides by divalent-ion additions. Journal of the American Ceramic Society. 1973;**56**:369-372. DOI: 10.1111/j.1151-2916.1973.tb12688.x

[2] Suszyńska M. Effect of impurity concentration and plastic deformation on dislocation density of KCl crystals. Kristall und Technik. 1974;**9**:1199-1207. DOI: 10.1002/crat.19740091015

[3] Kataoka T, Yamada T. Yield strength and dislocation mobility of KCl-KBr solid solution single crystals. Japanese Journal of Applied Physics. 1977;**16**:1119-1126. DOI: 10.1143/JJAP.16.1119

[4] Boyarskaya YS, Zhitaru RP, Palistrant NA. The anomalous behaviour of the doped NaCl crystals compressed at low temperatures. Crystal Research and Technology. 1990;**25**:1469-1473. DOI: 10.1002/crat.2170251219

[5] Boyarskaya YS, Zhitaru RP, Palistrant NA. Influence of the state of the impurity on the deformation-rate dependence of the yield stress of NaCl:Ca single crystals. Soviet Physics - Solid State. 1990;**32**:1989-1990

[6] Okazaki K. Solid-solution hardening and softening in binary iron alloys. Journal of Materials Science. 1996;**31**:1087-1099. DOI: 10.1007/BF00352911

[7] Tabachnikova ED, Podolskiy AV, Smirnov SN, Psaruk IA, Liao PK. Temperature dependent mechanical properties and thermal activation plasticity of nanocrystalline and coarse grained Ni-18.75 at.% Fe alloy. IOP Conference Series: Materials Science and Engineering. 2014;**63**:012105. DOI: 10.1088/1757-899X/63/1/012105

[8] Pratt PL, Harrison RP, Newey CWA. Dislocation mobility in ionic crystals. Discussions of the Faraday Society. 1964;**38**:211-217. DOI: 10.1039/DF9643800211

[9] Newey CWA, Harrison RP, Pratt PL. Precipitation hardening and dislocation locking in doped NaCl. Proceedings of the British Ceramic Society. 1966;**6**:305-316

[10] Chin GY, Van Uitert LG, Green ML, Zydzik G. Hardness, yield strength and Young's modulus in halide crystals. Scripta Metallurgica. 1972;**6**:475-480. DOI: 10.1016/0036-9748(72)90031-2

[11] Green ML, Zydzik G. Effect of heat treatment on the microhardness of some mixed and doped alkali halides. Scripta Metallurgica. 1972;**6**:991-994. DOI: 10.1016/0036-9748(72)90159-7

[12] Andreev GA, Klimov VA. Influence of the state of an impurity on the microhardness of NaCl:Sr single crystals. Soviet Physics - Solid State. 1980;**22**:2042-2043

[13] Buravleva MG, Rozenberg GK, Soifer LM, Chaikovskii EF. Changes in the flow stress of LiF:Mg^{2+} and LiF:Co^{2+} crystals during precipitation of solid solutions. Soviet Physics - Solid State. 1980;**22**:150-152

[14] Narasimha Reddy K, Subba Rao UV. Influence of gadolonium impurity on microhardness of host alkali halide crystal. Crystal Research and Technology. 1984;**19**:K73-K76. DOI: 10.1002/crat.2170190730

[15] Strunk H. Investigation of cross-slip events in NaCl crystals by transmission electron microscopy. Physica Status Solidi (A). 1975;**28**:119-126. DOI: 10.1002/pssa.2210280111

[16] Appel F, Messerschmidt U. The interaction between dislocations and point obstacles: A comparison of the

interaction parameter distributions obtained from computer simulation and from in situ high voltage electron microscopy straining experiments. Materials Science and Engineering: A. 1982;**52**:69-74. DOI: 10.1016/0025-5416(82)90070-2

[17] Messerschmidt U, Appel F, Schmid H. The radius of curvature of dislocation segments in MgO crystals stressed in the high-voltage electron microscope. Philosophical Magazine A. 1985;**51**:781-796. DOI: 10.1080/01418618508237587

[18] Kataoka T, Ohji H, Morishita H, Kishida K, Azuma K, Yamada T. In-situ observation of moving dislocations in KCl crystal by laser-light topography. Japanese Journal of Applied Physics. 1989;**28**:L697-L700. DOI: 10.1143/JJAP.28.L697

[19] Kataoka T, Ohji H, Kishida K, Azuma K, Yamada T. Direct observation of glide dislocations in a KCl crystal by the light scattering method. Applied Physics Letters. 1990;**56**:1317-1319. DOI: 10.1063/1.102504

[20] Kataoka T. The light scattering topography method: Direct observation of moving dislocations. Butsuri. 1992;**47**:713-716. (in Japanese)

[21] Messerschmidt U. Dislocation Dynamics during Plastic Deformation. Berlin Heidelberg: Springer; 2010. DOI: 10.1007/978-3-642-03177-9

[22] Indenbom VL, Chernov VM. Determination of characteristics for the interaction between point defects and dislocations from internal friction experiments. Physica Status Solidi A: Applications and Material Science. 1972;**14**:347-354. DOI: 10.1002/pssa.2210140142

[23] Schwarz RB, Granato AV. Measurement of the force-distance profile for the interaction between a dislocation and a point defect. Physical Review Letters. 1975;**34**:1174-1177. DOI: 10.1103/PhysRevLett.34.1174

[24] Ivanov VI, Lebedev AB, Kardashev BK, Nikanorov SP. Interaction of dislocations with pinning centers in magnesium at temperatures 295-4.2K. Soviet Physics - Solid State. 1986;**28**:867-868

[25] Kosugi T, Kino T. Experimental determination of the force-distance relation for the interaction between a dislocation and a solute atom. Journal of the Physical Society of Japan. 1987;**56**:999-1009. DOI: 10.1143/JPSJ.56.999

[26] Kosugi T. Temperature dependence of amplitude-dependent internal friction due to simultaneous breakaway of a dislocation from several pinning points. Materials Science and Engineering: A. 2001;**309-310**:203-206. DOI: 10.1016/S0921-5093(00)01792-5

[27] Gremaud G. Dislocation-point defect interactions. Materials Science Forum. 2001;**366-368**:178-246. DOI: 10.4028/www.scientific.net/MSF.366-368.178

[28] Dotsenko VI. Stress relaxation in crystals. Physica Status Solidi B. 1979;**93**:11-43. DOI: 10.1002/pssb.2220930102

[29] Urusovskaya AA, Petchenko AM, Mozgovoi VI. The influence of strain rate on stress relaxation. Physica Status Solidi (A). 1991;**125**:155-160. DOI: 10.1002/pssa.2211250112

[30] Johnston WG, Gilman JJ. Dislocation velocities, dislocation densities, and plastic flow in lithium fluoride crystals. Journal of Applied Physics. 1959;**30**:129-144. DOI: 10.1063/1.1735121

[31] Granato AV, Lücke K. Theory of mechanical damping due to dislocations. Journal of Applied Physics. 1956;**27**:583-593. DOI: 10.1063/1.1722436

[32] Blaha F, Langenecker B. Dehnung von Zink-kristallen unter ultraschalleinwirkung. Naturwissenschaften. 1955;**42**:556. DOI: 10.1007/BF00623773

[33] Ohgaku T, Takeuchi N. The relation of the Blaha effect with internal friction for alkali halide crystals. Physica Status Solidi A: Applications and Material Science. 1988;**105**:153-159. DOI: 10.1002/pssa.2211050115

[34] Ohgaku T, Takeuchi N. Relation between plastic deformation and the Blaha effect for alkali halide crystals. Physica Status Solidi A: Applications and Material Science. 1989;**111**:165-172. DOI: 10.1002/pssa.2211110117

[35] Ohgaku T, Takeuchi N. Interaction between a dislocation and monovalent impurities in KCl single crystals. Physica Status Solidi A: Applications and Material Science. 1992;**134**:397-404. DOI: 10.1002/pssa.2211340210

[36] Ohgaku T, Teraji H. Investigation of interaction between a dislocation and a Br^- ion in NaCl:Br^- single crystals. Physica Status Solidi A: Applications and Material Science. 2001;**187**:407-413

[37] Ohgaku T, Hashimoto K. Strain rate sensitivity of flow stress under superimposition of ultrasonic oscillatory stress during plastic deformation of RbCl doped with Br^- or I^-. Materials Science and Engineering A. 2005;**400-401**:401-404. DOI: 10.1016/j.msea.2005.03.058

[38] Ohgaku T, Matsunaga T. Interaction between dislocation and divalent impurity in KBr single crystals. IOP Conference Series: Materials Science and Engineering. 2009;**3**:012021. DOI: 10.1088/1757-899X/3/1/012021

[39] Kohzuki Y. Study on the interaction between a dislocation and impurities in KCl:Sr^{2+} single crystals by the Blaha effect-part IV influence of heat treatment on dislocation density. Journal of Materials Science. 2009;**44**:379-384. DOI: 10.1007/s10853-008-3150-8

[40] Kohzuki Y. Bending angle of dislocation pinned by an obstacle and the Friedel relation. Philosophical Magazine. 2010;**90**:2273-2287. DOI: 10.1080/14786431003636089

[41] Ohgaku T, Migiuma S, Nagahira D. Interaction between dislocation and defects induced by X-irradiation in alkali halide crystals. Radiation Measurements. 2011;**46**:1385-1388. DOI: 10.1016/j.radmeas.2011.05.067

[42] Nevill GE, Brotzen FR. The effect of vibrations on the static yield strength of low-carbon steel. Journal of ASTM International. 1957;**57**:751-758

[43] Langenecker B. Effects of ultrasound on deformation characteristics of metals. IEEE Transactions on Sonics and Ultrasonics. 1966;**SU-13**:1-8. DOI: 10.1109/T-SU.1966.29367

[44] Izumi O, Oyama K, Suzuki Y. Effects of superimposed ultrasonic vibration on compressive deformation of metals. Materials transactions, JIM. 1966;**7**:162-167. DOI: 10.2320/matertrans1960.7.162

[45] Evans AE, Smith AW, Waterhouse WJ, Sansome DH. Review of the application of ultrasonic vibrations to deforming metals. Ultrasonics. 1975;**13**:162-170

[46] Jimma T, Kasuga Y, Iwaki N, Miyazawa O, Mori E, Ito K, et al. An application of ultrasonic vibration to the deep drawing process. Journal of Materials Processing Technology. 1998;**80-81**:406-412. DOI: 10.1016/S0924-0136(98)00195-2

[47] Susan M, Bujoreanu LG. The metal-tool contact friction at the ultrasonic vibration drawing of ball-bearing steel

wires. Revista de Metalurgia (Madrid). 1999;**35**:379-383. DOI: 10.3989/revmetalm.1999.v35.i6.646

[48] Murakawa M, Jin M. The utility of radially and ultrasonically vibrated dies in the wire drawing process. Journal of Materials Processing Technology. 2001;**113**:81-86. DOI: 10.1016/S0924-0136(01)00635-5

[49] Susan M, Bujoreanu LG, Găluşcă DG, Munteanu C, Mantu M. On the drawing in ultrasonic field of metallic wires with high mechanical resistance. Journal of Optoelectronics and Advanced Materials. 2005;**7**:637-645

[50] Lucas M, Gachagan A, Cardoni A. Research applications and opportunities in power ultrasonics. Proceedings of the Institution of Mechanical Engineers, Part C. 2009;**223**:2949-2965. DOI: 10.1243/09544062JMES1671

[51] Siddiq A, El Sayed T. Ultrasonic-assisted manufacturing processes: Variational model and numerical simulations. Ultrasonics. 2012;**52**:521-529. DOI: 10.1016/j.ultras.2011.11.004

[52] Makhdum F, Phadnis VA, Roy A, Silberschmidt VV. Effect of ultrasonically-assisted drilling on carbon-fibre-reinforced plastics. Journal of Sound and Vibration. 2014;**333**:5939-5952. DOI: 10.1016/j.jsv.2014.05.042

[53] Graff KF. Ultrasonic metal forming: Processing. In: Gallego-Juarez JA, Graff KF, editors. Power Ultrasonics: Applications of High-Intensity Ultrasound. Cambridge UK: Elsevier; 2015. pp. 377-438

[54] Cottrell AH, Bilby BA. Dislocation theory of yielding and strain ageing of iron. Proceedings of the Physical Society of London. 1949;**62**:49-62. DOI: 10.1088/0370-1298/62/1/308

[55] Pamplin BR, editor. Introdction to crystal growth methods. In: Crystal Growth. Pergamon Press: Oxford; 1975. pp. 1-11

[56] Sirdeshmukh DB, Sirdeshmukh L, Subhadra KG. In: Hull R, Osgood RM, Sakaki H Jr, Zunger A, editors. Alkali Halides: A Handbook of Physical Properties. Berlin Heidelberg: Springer-Verlag; 2001. p. 41

[57] Conrad H. Thermally activated deformation of metals. Journal of Metals. 1964;**16**:582-588. DOI: 10.1007/BF03378292

[58] Kohzuki Y, Ohgaku T. Activation energy for dislocation breakaway from various monovalent ions in NaCl single crystals. Philosophical Magazine. 2016;(96):3109-3119. DOI: 10.1080/14786435.2016.1227099